Traveling the 38th Parallel

The publisher gratefully acknowledges the generous support of Gordon E. and Betty I. Moore as members of the Publisher's Circle of the University of California Press Foundation.

The publisher gratefully acknowledges the generous support of the General Endowment Fund of the University of California Press Foundation.

Traveling the 38th Parallel

A WATER LINE AROUND THE WORLD

DAVID CARLE AND JANET CARLE

UNIVERSITY OF CALIFORNIA PRESS
Berkeley   Los Angeles   London

University of California Press, one of the most distinguished
university presses in the United States, enriches lives around the
world by advancing scholarship in the humanities, social sciences,
and natural sciences. Its activities are supported by the UC Press
Foundation and by philanthropic contributions from individuals
and institutions. For more information, visit www.ucpress.edu.

University of California Press
Berkeley and Los Angeles, California

University of California Press, Ltd.
London, England

Library of Congress Cataloging-in-Publication Data
Carle, David, 1950-
    Traveling the 38th Parallel : a water line around the world /
David Carle, Janet Carle.
        p. cm.
    Includes bibliographical references and index.
    ISBN 978-0-520-26654-4 (cloth : alk. paper)
    1. Oceanography. 2. Hydrology. 3. Human geography. I. Carle,
Janet, 1953- II. Title.
    GC28.C37 2013
    551.48—dc23   2012030001

Manufactured in the United States of America

22 21 20 19 18 17 16 15 14 13
10 9 8 7 6 5 4 3 2 1

In keeping with a commitment to support environmentally responsible
and sustainable printing practices, UC Press has printed this book
on Rolland Enviro100, a 100% post-consumer fiber paper that is FSC
certified, deinked, processed chlorine-free, and manufactured with
renewable biogas energy. It is acid-free and EcoLogo certified.

This book is for Nick and Ryan, our sons,
and for the world they inherit,
and in memory of Larry Gibson,
Keeper of the Mountain, 1946–2012.

If nature can heal an injured land, it can heal our . . . souls as well. That's why saving Mono Lake is a matter of saving, and healing, ourselves. Let's not change the planet. Let the planet change us.

David Gaines, 1986

Contents

Introduction

You are undertaking the first experience, not of the place,
but of yourself in that place . . . for nobody can discover the
world for anybody else. It is only after we have discovered
it for ourselves that it becomes a common ground and a
common bond, and we cease to be alone.

Wendell Berry (1991, 43)

On June 25, 2010—exactly sixty years after the beginning of the Korean War—we stood at the edge of the Demilitarized Zone (DMZ), a 150-mile-long, 2.5-mile-wide strip of land on the 38th parallel that spans the Korean Peninsula. A soldier manning a nearby guard tower stared across the water, ready to strike at any threat from the north. An egret at the river stared into the water, ready to strike at fish. With people excluded from the DMZ for so many decades, the wetlands and forests behind barbed wire have become a haven for wildlife and an inadvertent experiment in ecological recovery.

That Korean experience was part of an around-the-world exploration we undertook along the 38th parallel, seeking water-related environmental and cultural intersections. Latitude lines are part of an abstract grid that humans have superimposed over the globe. Though imaginary, the lines coincide with physical realities dictated by this tilted planet's orbit around the sun. People who share a particular latitude, north or south of the equator, also share a specific orientation to the sky and experience the same day lengths during each annual cycle of seasons. Identical constellation patterns span the heavens every night at the same latitude around the world. The latitude and longitude grid helps travelers know where they are, and helps them navigate to destinations and recognize when they have arrived. Exploring a particular line can clarify both physical and cultural routes chosen by humanity, both historically and at this moment in time.

Thirty-eight degrees north (38°N) is a temperate, middle latitude where human societies have thrived since the beginning of civilization. Between extremes of climate found farther north and south, successful agriculture and the growth of cities along this latitude depended on many ingenious solutions to the problem of water supply. Altering natural water systems has consequences though, and, at this point in human history, environmental challenges associated with water are ubiquitous.

Our geographical exploration focused on stories of wetlands conservation; groundwater and river pollution; groundwater aquifers threatened by overdrafting for agriculture and domestic supplies; dams and aqueducts to store and transport water hundreds of miles to serve distant farms and cities; water limits generating tensions between cities, states, and nations; and the effects of climate change on the global water cycle. That grim list of challenges and problems was balanced by the inspiring work of scientists, environmental activists, and resource agency employees striving to renew what has been damaged, recover what has been lost, and preserve what remains pristine.

Water is the key to life on Earth, shaping the abundance and patterns of that life even when the influence is indirect. Lakes, rivers, and wetlands are particularly rich places of biodiversity. Adaptations that conserve water are the signature traits of desert plants and animals, while long-distance bird migrations are ultimately a quest to avoid water-freezing conditions, which threaten every water-rich living cell. As an erosion force, "water at work" explains many landforms. Civilizations have grown, evolved, or failed as they tapped water below ground, captured it behind dams, or moved it across the landscape to supply farmers and city dwellers.

Maps depict crosshairs where the 38°N latitude and 119°W longitude lines center on the oval expanse of Mono Lake, east of the Sierra Nevada crest, where we worked for two decades as park rangers and still live. The lake is a salty inland sea, too alkaline for fish, but with enormous quantities of algae, brine shrimp, and flies that feed millions of birds. Picturesque tufa towers decorate the shore where springs brought calcium into the carbonate-rich lake and formed limestone spires underwater. The towers were exposed as the lake surface dropped after the city of Los Angeles began diverting streams from the Mono Basin in 1941. A sixteen-year legal battle began in 1978 to save the shrinking, increasingly salty lake ecosystem, pitting concerned citizens and scientists against Los Angeles. In 1994 the David-versus-Goliath contest was won by little David. A small group of leaders with the Mono Lake Committee had built a national organization of supporters and volunteers that prevailed in court, ensuring that the giant city shares enough water with the lake to keep it alive and healthy. Such an outcome offers hope for a world facing many similarly daunting challenges.

The Mono Lake Tufa State Reserve was created in 1982 with a single ranger assigned to manage the new park. We, together, became that ranger, job-sharing until our retirement. It was a gratifying privilege to relate the lake's story to visitors from around the world, and water became the dominant subject of our ranger careers and post-retirement lives at Mono Lake. David has written several books about California water issues, while Janet still trains and coordinates the volunteers who conduct public and school tours at Mono Lake, explaining the lake's unique qualities and significant environmental history. Shaped by that background, our interest in circling the world took on a deeper meaning when we realized that the 38th parallel route intersects with so many lakes, rivers, and estuaries facing significant challenges. Our journey of discovery became a study in environmental geography.

The campaign to save Mono Lake succeeded due to a few key citizens and scientists who persevered, so we were particularly intrigued by the environmental protection campaigns we encountered during our travels. In every nation and state, we found dedicated activists, educators, land managers, and local experts willing to explain their difficulties and accomplishments. Passionate people were the rays of hope shining on problems around the world that could otherwise seem depressingly insurmountable.

Listing major cities is only one way to describe our route along the 38°N latitude: Seoul, the South Korea capital; Athens, Greece; and Córdoba, in southern Spain, make that list, along with Louisville, Kentucky; the Kansas cowtown of Dodge City; and San Francisco, California. Most of our traveling happened outside cities, though, away from well-beaten tourist tracks. The landscapes between those points provided conditions for our most interesting discoveries. The 38°N latitude bisects the Mediterranean Sea; coincides with much of the ancient Silk Road, which knit together oasis towns between China and Turkey along the edges of deserts; and intersects with the mouth of China's Yellow River, finding that nation's "Mother River" again and again as the channel meanders north and south of it. The line passes through the headwaters regions of the Tigris and Euphrates Rivers in Turkey and the Guadalquivir and Guadiana Rivers of Spain and Portugal. It finds Chesapeake Bay on the eastern edge of the United States, and the Colorado River in the West, after crossing the Great Plains along the Santa Fe Trail, the historic corridor that followed the Arkansas River upstream. The parallel also encounters islands where the surrounding sea shapes cultures and environments: the Azores, alone in the vast Atlantic; Sicily and several Greek Islands in the Mediterranean Sea; Japan's Honshu and Sado Islands; and the Farallones off the coast of California.

Within the United States, we learned about the decline of the Chesapeake Bay and San Francisco Bay ecosystems; mountaintop removal coal mining, which is filling in the creeks of West Virginia; wetlands in the heart of Kansas and the overdrafted Ogallala Aquifer; uranium mill tailings being removed from beside the Colorado River at Moab, Utah; and thirsty Las Vegas trying to tap into northern Nevada's groundwater, a quest reminiscent of the twentieth-century long-distance water-grab by Los Angeles in California.

On the route, we encountered many of the world's longest aqueducts that move water to cities from distant sources. One of the longest shifts river water 542 miles in Turkmenistan to regional farms and to Ashgabat, the national capital. Beijing is served by a 190-mile canal tapping reservoirs on the Yellow River, and the city plans to move additional water from the Yangtze River farther south. Athens, Greece, imports water from reservoirs 135 miles away. Spain's Tajo-Segura aqueduct extends from the wetter north to parched southern provinces where most of the nation's growth is happening, repeating a pattern familiar to Californians. In California, the Los Angeles Aqueduct extends 338 miles from the Mono Lake Basin; San Diego reaches for its water 444 miles into Northern California via the California State Water Project; and San Francisco's Hetch Hetchy System delivers Tuolumne River water 167 miles from the Sierra Nevada.

Flooding valleys to create reservoirs at the upper ends of these water systems has often been controversial. In Turkey, one of the world's oldest towns, Hasankeyf, is now threatened with inundation by a dam being built on the Tigris River. In California, the early-twentieth-century fight over whether to dam the Tuolumne River and flood Hetch Hetchy Valley inside Yosemite National Park helped clarify the meaning of national park status. Portugal's Alqueva Dam has been called a white elephant based on a long list of concerns about its costs and actual benefits. Water stored behind dams has not always been sufficient to deliver all the long-term benefits originally forecast, including energy generation, farm irrigation, and reservoir recreation. This was a lesson at the New Melones Dam on California's Stanislaus River. An unsuccessful battle against filling that reservoir pointed the environmental group Friends of the River toward many later successes.

Our route also brought us to natural lakes, inland seas, and coastal lagoons that have much in common with salty Mono Lake. Saline lakes are often underappreciated natural environments rather than "dead seas." Lagoons on the east coast of Spain support an ecosystem of shrimp, flies, and birds similar to that of Mono Lake, but add mud baths for people as a local attraction. *Tuz golu* means "salt lake" in Turkish, and at the namesake body of water, companies harvest salt while migratory birds nest and feed. Turkey's largest lake is Van, with alkaline water that forms towers similar to Mono's famous tufa. We encountered an ongoing national effort in central China to save another vast alkaline sea, Qinghai Lake, from the effects of sedimentation and stream diversions.

East of Qinghai, the city of Lanzhou was the historic junction for the Silk Road's northern and southern routes, over which traders carried goods between China and Turkey by moving between oasis towns. Water, or the lack of it, largely dictated that route's course and shaped the history of its cities and cultures. We traced the Silk Road (and its ubiquitous camels) across western China, through Turkmenistan, and all the way through Turkey.

In Korea, after venturing inside the DMZ, we learned about the renewal of an urban river through downtown Seoul and about the much more controversial Four Rivers Restoration Project, which is deepening and channelizing the banks of the Han River upstream from Seoul, carving away natural wetlands, and displacing rice farmers.

Wherever wetlands exist, wildlife populations are particularly rich. Wherever environmental manipulation damages natural systems, protecting and restoring habitat for endangered species becomes critical. We visited captive breeding programs for the Japanese Crested Ibis and for the Iberian Lynx in Spain and learned about endangered seals and sea turtles on Greek islands. Wetlands once thought lost forever were being restored in San Francisco Bay, the Sacramento–San Joaquin Delta, and in Point Reyes National Seashore to benefit beleaguered species of fish, birds, and amphibians.

At several places where water and land funnel migrating birds into narrow corridors, we joined with volunteers and scientists in annual counts of raptors and songbirds. Because locations on the 38th parallel are midway between the Arctic and the tropics, they experience four distinct seasons. Birds use north-south migration routes across latitudes to seek "endless summer," or at least avoid the harshest effects of winter. We joined bird counters at Point Reyes, near the Golden Gate on California's coast; at Chesapeake Bay on the U.S. East Coast; and in Sicily, where the count began as a campaign to end an annual slaughter of migrating raptors.

The woman behind the Sicilian campaign endured threats of violence as she built an international monitoring effort. Her inspiring success story was one of many we encountered—stories of dedication and struggle that enriched our journeys. The list of people who took time to meet with and educate us grew amazingly long. Acknowledgments in the back of this book include their names along with websites for the associated agencies, environmental groups, and campaigns that are striving to protect communities, lands, and local cultures.

North or south of the 38th parallel, few other global latitudes pass through so many states and nations. Because of land mass and population concentrations along the 38th, the Robinson map projection system, now widely used for world maps (and adopted by the National Geographic Society in 1988), selects 38° North and 38° South as "standard parallels"—the two lines running across the map along which size and shape are depicted most accurately, while distortions are greater at all other latitudes.

Following a straight line in almost any direction all the way around the world makes comparisons and connections between places more evident. Other latitudes would also provide meaningful insights, of course, but we encountered an intriguingly long list of stories within a half-degree either side of the 38°N line. We strayed farther afield only when travel logistics made it necessary or, in a few cases, to reach parts of an extended watershed that clarified the story of a river or lake along our route.

Although this narrative proceeds westward around the globe as though describing one continuous journey, we actually spread the traveling across three years in separate trips to Europe and Asia and across the United States. We traveled by bus, train, or rental car; flew between countries and across some of the long-distance segments within China; and trekked on foot, on bicycles, and by boat across California on a trip that put all of the global experiences into perspective.

This whole adventure grew out of an interest in geography that seems hardwired into most humans and—at the age we learn that the Earth is a round ball—often manifests itself in questions like "Where would I end up if I dug through to the other side of the world?" Many of us know the pull of the unknown and the road not (yet) taken, the attraction of mysterious unlabeled, "white" places on the map.

In recent years, Global Positioning System (GPS) devices have become popular navigation tools for hikers and drivers. The Google Earth website provides satellite imagery of the entire world referenced to latitude and longitude. Such widespread interest in navigation suggests that other people may want to visit the locations we describe, or may be inspired to explore other latitudes in similar detail. For this book, we record latitude and longitude coordinates in the classic degrees-minutes mode (38°30'N), rather than digital equivalents (38.5°N). Throughout our travels we posted updates at http://paralleluniverse38n.blogspot.com, with hundreds of additional color photographs and links to websites that may help people who want to follow in our footsteps.

Each travel segment was carefully planned, yet there were unanticipated encounters, a few pulse-raising adventures, and events that forced us to alter our itinerary. An ash cloud from a volcanic eruption in Iceland enveloped northern Europe and stopped air travel while we were on the Azores Islands. Nationwide strikes in Greece closed the airport the day we were to fly to Athens. The only buses going to our destinations in Turkey were sometimes fully booked due to school holidays. Planes were often late. Taxi drivers became lost. Kind strangers often helped us find our way. Successful travel is about serendipity, flexibility, and acceptance.

On the morning of March 11, 2011, four weeks before our planned trip to Asia, the Earth shook with incredible force. The magnitude 9.0 earthquake had its epicenter at 38°19'N, 142°23'E, about sixty miles offshore from Japan's largest island, Honshu. The tsunami that struck the east coast minutes after the initial quake caused the most tragic destruction. Entire coastal villages were demolished. Almost 20,000 people died and a half million were displaced. The quake damaged several nuclear power plants where reactors lost cooling capability and melted down, releasing radioactivity into the atmosphere and the sea. The nation struggled to recover and to address critical shortages of food, water, and medical supplies. It was not an appropriate time to visit the 38th parallel in Japan.

Six months into Japan's reconstruction of the area, we finally did visit and were able to write the final chapters of this book. Progress toward revival and recovery since the tragedy were exceptional and inspiring.

Indeed, Honshu was not the only place where we observed a desire to renew conditions for abundant life that have been diminished or destroyed. We found it nearly everywhere around the world on the 38th parallel.

Wendell Berry wrote that "nobody can discover the world for any-body else." We traveled the 38th parallel to put our feet in the Tigris and Euphrates, the rivers that served so much of early human history; to personally experience the dense press of humanity where China's Yellow River has nourished and tested its residents for thousands of years; to note the remnants of historic waterwheels and aqueducts and cisterns across southern Europe; to compare the chill air of Central Asian glacial valleys to the summer monsoon of South Korea; and feel the changes in weather and humidity as we left the verdant eastern states of our home nation and returned to the arid West. Our discoveries are shared here because our impressions, including the voices of people we met along the way, can be a first step toward your personal experience, toward a common bond that values a long-term, sustainable relationship with water around the world.

PART ONE Asia

What we have to work on in the twenty-first century is to overcome the division between people and the environment, so the future of humans won't be different from the future of nature.

South Korean high school student

Map 1.   Asia
1. Japan; 2. South Korea; 3. Xi'an; 4. Yinchuan; 5. Lanzhou; 6. Qinghai Lake; 7. Kashgar; 8. Turkmenistan; 9. Tigris River; 10. Euphrates River; 11. Cappadocia, Turkey

The Four Rivers "Restoration" Project

Our journey began in Korea. Hundreds of years ago, Seoul, the capital of modern South Korea, was a newly founded village along the banks of a picturesque creek called Chonggyecheon (37°35′N). As the city grew, the creek became a sewer and finally was covered over by concrete and a freeway. Several years ago, Mayor Lee Myung-bak decided to bring the creek back to the daylight. Now, a seminatural river parkway serves the urban residents of the city. We walked there from our lodge, a traditional inn with bamboo walls, sliding doorways, and an Ethernet-connected computer in every room; modern Seoul is one of the world's most wired nations.

Walkways line both banks of the flowing creek, which is punctuated by cascades and stepping-stone bridges. Chonggyecheon impressed us as an urban park, though water from the Han River must be pumped in at considerable energy costs to enhance the flow. The park is used day and night and clearly appreciated by local residents, from children splashing in the creek to romantic couples strolling hand in hand and elderly dog walkers. The mayor moved on to become the nation's president, and his new campaign, to redesign South Korea's four largest rivers—the Han, Nakdong, Geum, and Yeongsan—was a much more controversial objective.

In 2009, gigantic excavators began carving riverside bluffs away to double the width of the channels while digging them 12 to 18 feet deeper. The $20 billion national project encompassed the construction of sixteen new dams on the main channels of the four rivers, plus five more on their tributaries, the enlargement of eighty-seven existing small dams, and the addition of concrete lining along 200 miles of riverbank. President Lee called the effort part of a "Green New Deal" intended to store water against the prospect of drought, prevent flooding, improve water quality, restore river ecosystems, promote river-related recreation, and (perhaps above all) to stimulate the economy by creating 190,000 construction jobs and spending a sum equal to almost 20 percent of the nation's gross domestic product.

The goals of the Four Rivers Restoration Project sounded commendable, but when viewed closely, the list of benefits seemed exaggerated. We learned the details from and toured part of the construction site with the "Korean Federation for Environmental Movement" (KFEM). Their national headquarters is a comfortable old residence converted to offices, with commuter bicycles lining the walls. There we met with Choony Kim, the organization's chief of international affairs.

As Choony explained, South Korea, though densely populated, has plenty of water. Episodes of flooding occur primarily on upper tributaries rather than in the main river channels, where the work was focused. And the engineering approach will likely degrade, rather than improve, water quality, because slower water movement increases accumulations of algae and pollutants. Most important, the removal of natural wetlands and streamside vegetation clearly does not achieve "river ecosystem restoration," but rather destroys habitat. Natural beauty and wetlands critical to migratory birds and other wildlife were being replaced with bike paths, spraying fountains, and many miles of city-style parks.

When we pointed out that dams and channelization are old ideas about how to manage rivers, Choony said, "Many experts agree." She had worked for KFEM since 1995, after studying at the Yale School of Forestry and Environmental Studies in the United States. She spoke excellent English and was eager to talk to international visitors about the project, to counter publicity the government was providing to media outside Korea. Choony showed us dramatic before-and-after photographs taken at construction sites. As rivers are widened, villages and farms must be relocated. Her frustration showed as she explained that some of the displaced rice farmers, following years of encouragement by the government and environmental groups, had transitioned to organic farming, controlling insect pests in the paddies with ducks instead of pesticides and using weed-eating snails in place of herbicides. Shutting down those farms near the river meant 15 years of effort wasted.

An unexpected impact of losing so much farmland was a national "kimchi crisis" in the autumn after our visit, as a shortage of cabbage was blamed on lost farm production due to the Four Rivers Project. Kimchi, the spicy national dish, was served at every meal we ate in Korea. The cabbage shortage became a political issue, debated by the national legislature.

"The government calls this 'green economy,' but has no concern about the ecology," Choony said. "They just keep construction workers busy, busy, busy." The project is being pushed along aggressively because South Korea's presidents are limited to one five-year term and so President Lee "wants everything done in his time," Choony said.

The river widening and excavations do face opposition within the country. Polls find that the majority of the public opposes the work, both for its $20 billion cost and for its ecological impact. In a national election just before we visited, the president's ruling party had lost about half of its controlling majority in the legislature to opponents from other parties. Despite that change, the project was forging ahead, with a goal to finish by 2013. We asked Choony what KFEM hoped would happen to halt the work.

"The president cares about voices from the international community. He took part in the G20 meeting in Toronto, Canada [in 2010]. Korea will be the host government next time. He has created the Global Green Growth Institute. They want to advocate this Four Rivers Restoration Project to the international community, saying it creates jobs and 'green economic growth.' We say, 'No, if you look at this case, it's not real.' We give out information, but other nations listen to the government. So we need to give information to the international community, because he may listen to that voice. The international voice can come back to the domestic voice."

In the United States, a project of this type would require extensive environmental impact studies and would possibly be checked by lawsuits. In South Korea, KFEM lost a lawsuit against the project based on its impact to an endangered aster species (*Aster altaicus*) that depends on the riverbank environments. The court was persuaded by the government's promise to mitigate the impact by rescuing the asters from excavation areas and cultivating them in special plots.

From a watershed that gathers runoff from mountains to the east and southeast, the great Han River passes through Seoul, then turns northwest until it enters the ocean at the Demilitarized Zone. We followed the river upstream from Seoul to Yeoju, a city on the south fork of the Han. Our driver, KFEM photographer Jong-Hak Park, had a medallion hung on the dashboard of his car showing that he supported the International Crane Foundation. Now 68 years old, he had worked with KFEM for 12 years. "I love KFEM, because my grandson has allergy," he told us. "I heard that it's from environmental pollution, so I will keep the environment."

Two other young and energetic KFEM staffers, Yong-un Ma, the organization's wetlands expert, and Naree Jeong, coordinator of a Nature Conservation Team, showed us a six-mile stretch of the river where construction crews were building three dams and excavators were filling trucks with loads of riverbed sand and gravel, adding to mountains of material piled nearby. The scale of the work was shocking. We had never seen so many gigantic excavators at work, digging, widening, chewing up, and spitting out the former riverbed. The dams we saw under con-

Figure 1. Excavators widen and deepen the channel of the Han River as part of South Korea's controversial Four Rivers Restoration Project.

struction were close enough together so that a series of excavated "water basins," separated by gates and locks, could conceivably become a continuous canal to serve shipping. Opponents of the project think that is exactly the point, as a grand canal connecting the nation's rivers was one of President Lee's key campaign promises until negative public opinion stopped the plan.

The Koreans working against the project were inspiring. It seemed hopeless, yet they were bearing witness to anyone who would listen, and asking us to "please tell the rest of the world what is happening to our rivers." KFEM was created when several South Korean groups involved in anti-pollution and anti-nuclear issues merged in 1987. The nongovernmental organization has fifty local chapters, whose thirty thousand members pay monthly dues.

We were treated grandly that evening with a feast held at a restaurant overlooking the river owned by Hang Jin Lee, the head of the Yeoju chapter of KFEM. He told us that he is "not a born environmentalist, but about 10 years ago my friends asked me to work for the environment. I began to realize that environmental protection is very, very important. And now I cannot leave this work." Having given considerable thought to two opposing characteristics of humanity, consciousness and greed, he hoped that "we can regain the essence of consciousness—the relationship with living things and nonliving things."

From the window in our hotel room that night, we had a view of the Han River midway between the two work sites we had seen that day, where excavators continued to chew away at the riverbank.

The United States went through a river damming, channel straightening, and concrete armoring stage in the past century. We learned that a more effective way to manage watershed systems is to allow floodplains to absorb high-water episodes and regulate flows while the riverbank vegetation filters impurities. So it was frustrating to watch the living river bottom being pulled out and lush riverside plateaus being excavated. Although the momentum of the work seemed impossible to stop, opponents of the project remained determined. On a poster in the Yeoju KFEM office, the endangered aster plant was depicted restraining an excavator. Mr. Ma told us the words on the poster read: "Be Persistent; It Is Your River."

Ecological Recovery behind Barbed Wire

On the sixtieth anniversary of the start of the Korean War, June 25, 2010, we entered the Demilitarized Zone (DMZ) that separates North and South Korea. We were with a delegation from Pasadena, California, which has a "Friendship City" relationship with Paju, a booming South Korean city on the edge of the DMZ (37°46′N). Few travelers are allowed inside the Joint Security Area of the DMZ, administered by the United Nations, so this was a unique opportunity for our 38th parallel exploration.

A strip of land 150 miles long and 2.5 miles wide, the DMZ crosses the Korean Peninsula at the 38th parallel. A peace treaty was never formalized to end the Korean War, and North and South Korea each keep hundreds of thousands of soldiers massed along their borders in a tense military standoff, with the DMZ functioning as a no-man's-land between

Figure 2. Looking at North Korea across the Imjin River, South Korean guards keep watch on the Demilitarized Zone.

the two nations. The war began in 1950, the year David was born, and the cease-fire came in 1953, Janet's birth year. During our entire lives, soldiers have stood watch over a land fringed with barbed wire and perhaps a million landmines; at the same time, the land has healed itself and an inadvertent ecological recovery has occurred.

Most of that landscape was farmed for thousands of years, then ravaged when it became a battlefield. Ironically, after years of war and a peaceful standoff that has kept human activity out of the DMZ, eleven hundred types of plants now grow behind the barbed wire, and the land is green and lush. At the western edge of the DMZ, the Han and Imjin Rivers Estuary provides rich tidelands and riparian habitat for wildlife, including eighteen endemic fish species (found only there) and the endangered Korean Golden Frog (*Pelophylax chosenicus*). The Han River Estuary is the only one in South Korea that retains its natural function as a nursery for marine life in the Yellow Sea; others along the nation's coast have been degraded by wetlands reclamation or dams at river mouths.

Many endangered or threatened bird and mammal species are thriving in the estuary and in other wetland basins inside the DMZ, including three species of cranes that arrive each winter. Asiatic Black Bears (*Ursus thibetanus*) are found, as well as leopards, lynx, and possibly the exceedingly rare Siberian Tiger (*Panthera tigris altaica*), if glimpses seen on military surveillance cameras and territorial scratch marks on trees have been correctly interpreted.

Winter is the peak season for migratory birds; but the summer residents we saw included herons, egrets, and magpies; flocks of Bean Geese (*Anser fabalis*); and the colorful Black-Capped Kingfisher (*Halcyon pileata*). We had numerous fleeting looks at Water Deer (*Hydropotes inermis*), whose males have large tusks at their jaws, but no antlers.

Seeing any wildlife was a treat, as we were allowed only to travel directly to Daesong-dong village, where 212 South Koreans live inside the DMZ (37°57′N). The Pasadena delegation was granted special permission to enter the Zone to dedicate a new dental clinic there on the war's sixtieth anniversary. The group included Yung Nam, a Korean-American dentist, who was the inspiring force behind the clinic. Until this time, we had no idea that *any* people actually lived inside the DMZ, but both South and North Korea are allowed small "unification" or "peace" villages. The rest of the Zone remains unoccupied, and that is where the ecological experiment in human exclusion has been under way.

Our group passed through checkpoints where serious-looking soldiers closely examined our passports and paperwork before giving the driver a piece of cloth to hang from the window of our van—a low-tech signal that we had permission to be inside the Joint Security Area itself. Beyond the checkpoint, our initial impressions were of verdant forest and quiet. The energy and bustle of modern South Korea had been left behind.

Dr. Nam's plans to dedicate the dental clinic at Daesong-dong were thwarted at the last minute because of tension with North Korea over the torpedoing of a South Korean gunboat. United Nations administrators were concerned that North Korea might consider a ceremonial gathering, even for such a benevolent purpose, provocative. It was a reminder of just how volatile (and sometimes illogical) the relationship remains in that region. Although the clinic dedication was off, our delegation was allowed to visit the school where it will someday be established. From a rooftop observatory in Daesong-dong, our eyes were inevitably drawn toward structures in the North Korean "peace village" just 400 yards away. A North Korean flag flew from a provocatively tall pole, an invitation to a flagpole "arms race" that the South has declined.

Our guides pointed toward hillsides denuded by the fuel-wood scavenging of impoverished North Koreans, where the deforestation has caused severe floods and soil erosion. The southern side's thick forests offered a striking contrast, though Dr. Nam recalled barren hills around his home near Paju when he was growing up, during and just following the Korean War. After the war, the South Korean government put many citizens to work planting trees to reestablish forests, though the trees most often planted were conifers instead of native trees, so that habitats for many species never recovered.

Our delegation was lodged in Paju City. Municipal officials gave us business cards bearing the slogan (also mounted on every taxi in the city) "GG Paju," standing for "Good and Great Paju." Undeterred by the tensions of the DMZ that it borders, Paju is one of the fastest-growing cities in fast-growing South Korea. Paju's population had increased by over 80 percent in the prior decade, and the mayor's office had rather incredible plans to double the population in the coming decade by attracting high-tech industries. This frenzied economic activity presents a concern for people watching the ecological recovery within the neighboring DMZ and wondering what the future will hold, if and when that land is once again opened to development. Plans have been prepared for developing the Han River Estuary into a major seaport and building a new "Reunification City" in the Cheorwon Basin, one of the important wetland areas in the DMZ.

To promote transformation of the DMZ, a symbol of war, into "a place of peace among humans and between humans and nature," the DMZ Forum was founded in 1994, with headquarters in New York. Its members hope to build enough international support to create a nature preserve and international peace park and forestall any destructive consequences if the Zone ever opens to the world.

The local DMZ Ecological Research Institute works with South Korean schoolchildren to shape eco-friendly attitudes within the nation itself by taking students into the Civilian Control Zone (CCZ), a strip of land three miles wide that parallels the southern edge of the DMZ. Though also fringed with barbed wire and closely guarded, the CCZ is not off-limits to South Koreans. Some commute there to work on family farms, while many join international tourists visiting Korean War historic sites. Our guide was JungRok An, a student at the Korea University in Seoul who founded the Youth Exploration Team program for the institute, which brings high school students to the CCZ from all across Korea.

Only 20 years old, JungRok is an articulate and knowledgeable researcher, a dedicated proponent of environmental education, and a born leader. "To protect the DMZ for the future," he told us, "I think youth are the most effective." "EcoYouth" students receive one year of classroom orientation and attend field camps to learn research techniques, then begin individual or small-group ecological study projects within the CCZ. JungRok's own involvement had begun five years earlier with research on Swan Geese (*Anser cygnoides*). His father heads the institute's education section.

We were with a group of Youth Exploration Team students from high schools in Seoul on their first day of orientation. The institute's education staff handled two other groups that same day: high school students who were a full year into their studies, plus an elementary school class. Although South Koreans can visit the Civilian Control Zone, these students had never been there. Our itinerary included the Dorasan railroad station, where South Korean trains must stop, though the tracks continue to the north. Map displays explain how ties to Russia and Europe will become possible whenever the route reopens. We also went to places tourists never see, including wetlands along the Imjin River.

Along the Han and Imjin Rivers, local farmers from Paju grow rice and ginseng. Loosely strung electric lines serve irrigation pumps that lift river water to the fields. Ginseng has become a key moneymaker for farmers, but requires plastic covers on frames to shade the plants. The covers also keep out birds and other wildlife, so the continuing expansion of ginseng fields concerns JungRok. Rice fields, by contrast, can be beneficial feeding and resting wetlands for migrating waterfowl (a role we were familiar with from the Sacramento Valley and Delta in California, where the stubble left in flooded rice fields after harvests serves seasonal bird migrations). Along roadsides, dozens of red triangular signs with skulls and crossbones warned of land mines. The kids found the grim warnings fascinating and took photos of each other posed in front of the signs. At one point, as we got out of our van to look at a pond, soldiers on patrol told us we had to move along, that we were not supposed to stop there. Reminders of armed hostility were unavoidable.

That afternoon a drizzling rain began to fall. The annual summer monsoon was forecast to start that day, and it did, at 4 P.M., right on schedule. The highway back toward Paju paralleled the river estuary, where egrets fished behind a chain-link fence topped by razor wire.

We asked JungRok what he most wanted people to know about his project and the institute. "Before I started these DMZ ecological exploration teams, I thought, 'What can I do to help this environment that is now endangered by humans?' One way is to change the mind itself, so that's why I thought of youth. The whole environmental problem is on us, the youth, because it will happen when we are older, and it will happen for our children. So if youth are talking to adults about this environment and how it is beautiful and mystical and important, it could bring change."

On the van ride out of the CCZ, we rode with six high school students who had already completed a year of work. Armed with nets and collecting jars, they had spent the day sampling water quality in ponds and documenting populations of aquatic invertebrates and amphibians. They were eager to speak into our tape recorder, as if it were a direct audio link to the world, to explain what they wished people to know about the DMZ and their hopes for the future.

"Hello listeners," one girl began. Janet told her that she would someday be a television personality or a diplomat, which made all of the students laugh. "I hope you know that the DMZ in Korea is a very valuable, ecologically important place. Even though the 'demilitarized' name is a meaning about military values, there are also environmentally important things in there, and even after Korea is unified, we need to keep up the habitat interests of the place. Thank you."

A quiet young man from CheongShim International Academy provided this eloquent conclusion: "I think you two just realized, today, what I realized working as a DMZ member. You mentioned the irony you found here, how the war between people led to a better future for this environment. I think that's the main point. What we have to work on in the twenty-first century is to overcome that irony, the division between people and the environment, so the future of humans won't be different from the future of nature."

On September 3, 2010, the South Korean Ministry of the Environment announced it was designating three areas inside the DMZ, including the Chopyeongdo and Jangdanbando wetlands at the Imjin River Estuary, as "wetland protection areas." This classification was done to counter development plans being considered by "various government departments and local authorities . . . despite [the DMZ's] high ecological and scientific value" (Jong-hwi, 2010). Domestic legal designation as protected areas is the first step toward classification as national parks and UNESCO Biosphere Reserves.

China's Yellow River Delta

High-speed freeways paralleled the east coast on the long drive from Beijing, China, south toward the city of Dong Ying. Orchards were blooming on this mid-April day, and rows of cottonwood trees along the highway were leafing out, recently planted as part of a massive national campaign to counter climate change by capturing atmospheric carbon and, as we would see in central and western China, to erect a "green wall" against expanding deserts. Many millions of trees had already been planted.

Our guide and translator across eastern China was Kinder (pronounced "kin-dur") Jinde Shu, who has a special interest in natural places and migratory birds and could offer us helpful connections with China's parks and nature reserves. He was China's first member of the National Association for Interpretation (NAI), an American organiza-

Figure 3.   China's Yellow River gets its namesake color from the heavy load of silt it carries.

tion for professionals who conduct interpretive programs and design educational exhibits for parks. He has translated NAI materials into Chinese and coordinated training for China's parks and reserve staff. Our starting point on the 38th parallel in China would be the National Nature Reserve at the Yellow River Delta, where a special tour had been arranged by the deputy director, a personal friend of Kinder's.

The highway crossed over water channels along an almost perfectly flat landscape. Such terrain, along with the extremely high silt loads the Yellow River carries, explains why the river (the Huang Ho, in Mandarin Chinese) repeatedly fills its channel with silt and then jumps into new routes to the Bohai Sea, a northern gulf on the Yellow Sea. The mouth of the Yellow has shifted more than fifty times in the last 150 years. Today it enters the ocean several miles south of the 38th parallel, but just a few years ago it occupied a channel exactly on the latitude line.

This river carries the highest silt content in the world, a brownish-yellow sediment load that explains its name (*huang* means "yellow") and averages more than one billion tons of silt each year. "When the Yellow River flows clear" is a Chinese saying for an event that will never happen (similar to the English expression "when pigs fly.") The delta zone adds an average of 8,000 acres annually, and China's youngest wetlands keep expanding.

The Yellow River flows west to east, beginning as snowmelt in the Kunlun Mountains, crossing the Qinghai-Tibet Plateau, and passing through nine Chinese provinces. The 3,400-mile-long river's main stream and thirty-five tributaries have a combined length of 8,380 miles and drain a 290,500-square-mile watershed. It is China's second-longest river (the Yangtze, about 500 miles longer, follows a parallel path farther south). Our four-week trip across China began at the delta and brought us to the Yellow River several times more, far inland.

From our perspective traveling upriver from the coast, the river channel bent southwest through alluvial plains of windblown silt ("loess"), then turned north for hundreds of miles (circling the Ordos Desert within Inner Mongolia), before making a great sweep south again, then bending southwest toward its Tibetan headwaters. In central China we would follow the main course of the river for about 250 miles and then travel along a tributary for 100 miles, the Huang Shui, which descends from eastern Qinghai Province.

Where the Yellow River crosses flat floodplains, it has been confined between levees for thousands of years. Sediment inexorably settles onto the riverbed, forcing residents to keep elevating flood-protection walls in response. In places, the river runs 20 to 30 feet *above* adjacent farmlands and villages, putting 90 million people in the floodplain at tremendous risk if levees fail. Not only is the Yellow China's "Mother River," sustaining life along its length, but it is also the nation's "River of Sorrow." Archaeological evidence of disastrous floods dates back more than 2,000 years. In 1887, one of many historic Yellow River floods killed nearly 2 million people. The worst flood disaster in world history occurred here in August 1931, killing about 3.7 million people. Such numbers are worth a moment's reflection; China's vast landscape and population size make the scale of catastrophes unimaginable.

The Mother River has, ironically, also been given very little respect, as it has become a convenient sewer and waste disposal point for towns and factories. Some segments of the river are so badly polluted that the toxic water is even unfit for farm irrigation. Thirty percent of the river's fish species were extinct by 2007, according to the Chinese government.

From the bustling industrial city of Dong Ying, we drove about 40 miles to the delta. When we started out, we could scarcely believe a protected wetland would be at the end of that road. Construction zones gave way to farmland, then to China's second-largest oilfield, and finally to marshes. At the gate of the Yellow River Delta National Nature Reserve, a massive Oriental White Stork (*Ciconia ciconia boyciana*) lifted off in a flash of black and white. Twenty-eight pairs of storks had bred at the delta that year, which serves as an avian international airport each autumn, winter, and spring. Common Terns (*Sterna hirundo*) streaked by in a high wind as we walked down boardwalks with intricately carved handrails to elevated bird-watching towers.

Our reserve guide was Shen Kai, a young engineer on the staff and a bird expert who carried a camera with a huge telephoto lens. His supercharged energy kept us zooming between spots in the wetlands as we struggled to keep his car in sight up ahead. The nature reserve was created in 1992 to enhance and protect the delta and its wetlands for migratory and nesting birds. More than 6 million birds use the protected lands every year—296 species, including six of the world's nine species of cranes. Seven of the species that depend on the reserve's wetlands are "class 1 priority" species in China (equivalent to the United States' endangered species designation), including cranes, eagles, and storks.

Two separate sections of land, together totaling 378,000 acres, reflect the most recent major shift in the river's path: a northern parcel of the reserve coincides with the Yellow River's sea entrance before 1976 (on the 38th parallel), while the present channel passes through a larger parcel about 15 miles to the south. Huge oil and natural gas fields lie between the two areas, with some oil wells pumping inside the reserve itself. Pollution is a concern, along with habitat fragmentation caused by roads and installations, but the managers' greatest struggle is to secure enough "eco-water" from the main river to optimize ecological conditions in the marshes.

With increasing water consumption upstream, the Yellow River reached the sea only one year in the 1990s. As with our nation's Colorado River, which dries up before reaching the Gulf of California in Mexico, there have been too many competing demands for human use of Yellow River water. The national Yellow River Conservancy Commission (YRCC) responded to that problem by regulating water use, setting specific allotments for each of the nine provinces along the river basin. Amounts are based on a 1987 estimate that the river carries 15 trillion gallons of water annually, but—in another parallel to the Colorado River, whose flow rates were overestimated when states negotiated water rights—in the past decade the Yellow River's volume has fallen to 14 trillion gallons; in 2003 flows were below 12 trillion gallons. We visited during the spring dry season, yet flowing water was reaching the sea, thanks to the YRCC's new conservation regulations. Since 2003, there has been enough reliable flow to allow water diversions into wetland ponds, "but still only half of what that landscape needs," according to Shen Kai.

The rainy season here arrives with the hot days of the summer monsoon. Flow in the Yellow River peaks between July and October, elevated by heavy monsoonal rains across the eastern half of China. The summer months, spanning May and October, also bring more than 150,000 tourists. Spring is dry and windy, as we can attest: morning winds nearly swept us off our feet while we stood on the rooftop viewing platform of a new visitor center.

The modernistic building, completed in 2010, looms over the landscape, six stories high. It is called the "Ecological Port" because it sits on the south bank of the river where boat tours begin. From the rooftop we had jaw-dropping views of the marsh and the Yellow River heading toward the ocean, just a quarter-mile away. The new center did not yet offer any services—all funding seemed to have gone into the incredible structure—but it appeared large enough to handle thousands of visitors. On less windy days, two tour boats were available for cruising the river out into the nearby sea. A dramatic mixing line between blue ocean and yellow river water is a must-see that the wind denied us.

The howling gusts, thankfully, stopped during our lunch in the nearby town of Xian He (the name meaning "holy crane"). We were honored with a special restaurant meal arranged by Kinder's friend Li Juanzhang, the deputy director of the reserve. Mr. Li was unable to join us, but we feasted on dishes unfamiliar to us—sea cucumber and octopus. It was far too much food for us and our guides to finish, but as in Korea, Chinese culture calls for excess when honoring guests.

During lunch, Shen Kai told us that what he liked best about working at the reserve was protecting the birds and preserving the special character of the wetlands and tidal lands. He grew up in the local area and had a university degree in petroleum engineering.

That afternoon, with the air blessedly calm, we watched Red-Crowned Cranes (*Grus japonensis*) in flight. We had missed the breeding season for cranes, and most of them had migrated north again, but some were kept in cages and turned loose daily to fly free. Joining a group of Chinese tourists, we watched enraptured as the birds circled near us again and again. Cranes are the "birds of heaven," the "holy bird" of China, associated with love, happiness, and all good things, especially longevity. Their lifespans approach eight decades (the International Crane Foundation in Baraboo, Wisconsin, has documented an 83-year-old Siberian Crane). Cranes are often depicted together with pine trees in Chinese art as both are long-lived. Reverence for cranes is part of Chinese culture.

When the practice of releasing captive birds to fly free began, some birds, inevitably, chose not to return, despite the lure of abundant food. Mr. Li told us that was fine. "It was best for them. But when they do come back, it shows good relations between cranes and people."

That philosophical opinion was expressed during our evening dinner, another feast hosted by the reserve. Kinder had met the deputy director when they worked together on a Siberian Crane wetlands project. Mr. Li had begun working there twenty years earlier when the first protective status began on the delta as a local effort. Within a few years, in 1992, the reserve became a National Nature Reserve.

Beside each dinner plate that evening was a small toasting glass that a waitress skillfully kept full to the very brim with a rice liquor. There were many toasts, including one to adequate "eco-water" for the reserve. As Kinder translated, Mr. Li told us how difficult water management can be, but he added that serving the birds was a worthwhile, satisfying goal.

"From the view of conservation management, it is a challenge," he said. "Personally, it is a very heavy burden to maintain this piece of land, but it is an important job. Compared with the United States and other developed nations, where it is easier to preserve nature with established policies and strong public involvement, we have much to improve. Fortunately we have a supportive local government, so that gives us a lighter burden." For example, after the central government appropriated $5 million for wetlands restoration, state and local governments provided another $12 million. The regional wealth based on the oil and natural gas industries made that generosity possible.

The Colorado and Yellow Rivers have much in common: both drain vast regions within each country, both were named for the colors given by heavy loads of silt (*colorado* means "red" in Spanish), and so much water is diverted that both rivers stopped reaching the sea. Given the rivers' similarities, we agreed that there should be a "sister rivers" relationship established between the two. Although ecological health requires more water than the reserve has yet been able to secure, it was good to see that the notoriously dry Yellow River Delta had, at least, been addressed. Regulations imposed by the YRCC the entire length of the watershed show that China can respond to the conflicts between rapid industrialization, historic water uses by cities and farms, and the needs of the natural environment. How well national regulations are actually being implemented in the distant provinces was a question we would explore further as we crossed the country. We left the Yellow River Delta impressed by the enthusiasm of the reserve staff and hopeful that China's Mother River might continue to reach the sea with "birds of heaven" winging overhead.

The South-North Water Transfer Project

As we drove back toward Beijing, our route crossed the path of a new aqueduct being built to bring water from the Yangtze River to Beijing, but it was underground, out of sight. The first phases of the gigantic South-North Water Transfer Project were being constructed to shift water from southern China, where 80 percent of the nation's precipitation falls, to the drier north, particularly to address the thirst generated by industry and population growth near Beijing. While the imbalance between wet regions and human thirst is geographically reversed in California, where north-to-south aqueducts serve Southern California cities and farms, the approach to water supply issues is similar: build long-distance aqueducts. Ultimately, China intends to complete 1,800 miles of canals along three routes to move almost 12 trillion gallons of water each year. Work began in 2002 on the middle route from the

Han River, and the eastern route from the Yangtze. Construction crews missed a goal to finish the eastern aqueduct before China hosted the 2008 Summer Olympics; completion is now projected for 2013. The third, western route, which will begin high on the Tibetan Plateau at the headwaters of the Yangtze, faces even more daunting engineering challenges, so work there has been delayed.

The route of the eastern aqueduct uses existing parts of China's ancient Grand Canal. That feat of early engineering began 2,000 years ago and was continued by successive dynasties to make barge travel possible along 1,100 miles of canal connecting the Yangtze, Yellow, and other rivers. (Hangzhou, in the south, was finally linked with Beijing in 600 A.D.). Perhaps 2.5 million laborers worked on the canal and a paved highway along its length. The twenty-first-century construction is all about water supply, rather than transportation.

Water coming from the Yangtze River is highly polluted, necessitating construction of hundreds of costly sewage treatment plants in the north. Industrial pollution and China's national demand for water increased in step with the country's recent phenomenal growth as new industries, especially power plants, required more of the nation's already limited water supply. The Chinese have pioneered coal-fired power plant technologies that use much less water, yet overall, thirst still grows while the available water supply has actually declined.

Climate warming is altering rainfall patterns and melting glaciers in the Qinghai-Tibet Plateau, the headwaters for half of the Yellow River's flow. Thousands of lakes have dried up in the plateau, sometimes called China's water tower, and in 2007, Chinese scientists documented a 10 percent shrinkage in wetlands that feed the nation's two largest rivers. A series of reports sponsored by Circle of Blue and titled "Choke Point: China" that analyzed conflicts between water and energy noted: "China's total water resource, according to the National Bureau of Statistics, has dropped 13 percent since the start of the century. . . . That's as much water lost to China each year as flows through the mouth of the Mississippi River in nine months" (Schneider, 2011).

China's reliance on coal as a power source helped push it past the United States as the number-one emitter of greenhouse gases. In the past five decades, climate warming accelerated globally as the human population rose from 3 billion in 1960 to 7 billion in 2012. While the atmosphere's greenhouse effect is a natural phenomenon that is critical to life on Earth, humanity's burning of fossil fuels since the Industrial Revolution has elevated atmospheric levels of carbon dioxide and other greenhouse gases to levels not seen in the past 650,000 years. The carbon dioxide level, which was about 270 parts per million in 1800, rose past 390 parts per million in 2012. After analyzing all the variables that naturally alter climate, including fluctuations in solar radiation, variations in the Earth's rotation, and volcanic eruptions, computer models can match the observed trends only when they include the elevated greenhouse gases released by human burning of fossil fuels.

We hear so much about carbon dioxide and other greenhouse gases, we tend to forget that the effects of climate warming on the planet's life are felt, ultimately, through the water cycle. Energy trapped by an enhanced greenhouse effect translates into daily weather and long-term climate changes, because warmer oceans and land surfaces evaporate more water, and warmer air holds not only more energy but also more water vapor. Increasingly extreme and variable weather brings hurricanes, floods, and droughts. Annual snowpack and rainfall totals change, glacial melting accelerates, sea levels rise, and new runoff patterns complicate the management of reservoir storage for farm irrigation, industrial use, and drinking water. Across Asia, Europe, and the United States along the 38th parallel, we encountered examples of such challenges and learned how nations are responding.

We were returning to Beijing to catch a flight the next evening. Janet had visited that city before and thought the air seemed cleaner than the last time she was there, in the summer of 2007. Air pollution was so bad then that she could not see buildings a few blocks away. The difference might simply have been seasonal; we were traveling in the spring instead of summer. But major efforts had been made to clean up the air for the 2008 Olympic Games and Kinder told us that several polluting factories and power plants, shut down that year, had never reopened. Still, small particulates from coal-fired power plants, truck and car exhausts, and dust remain major problems in China's cities.

After walking on the streets in each city where we stayed, our throats and eyes would be irritated.

China is trying to address the air quality and climate-change problems with cutting-edge technologies to reduce coal plant emissions, with major investments in wind and solar energy, and by encouraging cleaner-running vehicles. Several of the cars and vans we traveled in were fueled with compressed natural gas, a cleaner-burning fuel than gasoline. Small three-wheeled electric trucks and electric motorcycles were common; we spotted a number of motorcycles parked on sidewalks with extension cords running inside stores to charge batteries. The national tree-planting campaign was another way to clean the air and reduce dust pollution.

Kinder had worked on an International Friendship Forest in conjunction with the Olympics. During our layover in the city, he was eager to show us the native plantings in the shadow of the Great Wall, north of Beijing, complete with an extensive trail system and interpretative signs. Unfortunately, a guard was keeping people off the trail that day due to fire danger, and the path's starting point had been commandeered as a motor scooter parking lot. Dust and sun bleaching made his signs hard to read. Kinder was obviously disappointed with the upkeep. This pattern was repeated at other national parks we saw in China: money and thought poured into a facility without long-term operational support. In most parks we saw grandiose buildings, fancy entrance gates and fountains, but too few maps, brochures, or information panels for visitors.

Kinder may be the person to correct that situation. Believing that it is important for Chinese national parks and nature reserves to provide meaningful scientific information and the conservation stories behind the scenery, he hopes to organize more training for park staff, including international exchanges with park interpreters.

From Beijing, we flew southwest to the transportation hub of Xi'an, where we took a rainy-day tour of the city's old walls and joined tourist hordes to view the famous Terracotta Warriors of Qin Shi Huang. Xi'an and the Great Wall near Beijing were two places in China where we did not stand out as the only Westerners.

Xi'an was once the eastern starting point for the historic Silk Road, a trade route we would follow for hundreds of miles in western China. It was the capital of the Han Dynasty during the peak of Silk Road trading. The ancient city is on the Wei River about 60 miles from where it joins the Yellow River, and about 110 miles upstream from Sanmenxia Dam. Kinder made sure we knew that the infamous dam had been designed by Soviet engineers, not Chinese.

Construction of the dam began in 1954, on a 900-foot-tall plug across the Yellow River at the narrows of the Three Gate Gorge—not to be confused with the Three Gorges Dam on the Yangtze River. More than 400,000 people were forced to relocate when Sanmenxia was built (more than a million lost their homes because of the Three Gorges Dam). As we had seen in South Korea along the Han River, forced resettlement is one of the drawbacks balanced against benefits of river damming to serve broader goals, despite the cost to local populations and to communities that must absorb those who are relocated. Although payments or land-for-land exchanges occur, often high-yield farmland in floodplains has been replaced with poorer-quality land on upper reaches.

The Sanmenxia Dam was supposed to control floods and trap the river's sediments, an explicit goal being "a clear Yellow River" below the dam in order to halt accumulations of sediments that led to floods whenever the river jumped course. Some Chinese experts warned about the folly of purposely trapping so much sediment. In 1963, just three years after the dam began storing water and sediment, the reservoir had accumulated four billion tons of material and lost 40 percent of its storage capacity. Hydroelectric turbines became clogged with silt. A "tail" of sediment was backing upriver, creating new threats of flooding as far upstream as Xi'an. The dam's gates had to be left open and turbines pulled out so water could pass through and flush some of the accumulation. The reservoir became nearly worthless for storing water or generating electricity. In 2000 a second large dam was completed a bit farther downstream on the Yellow River. Although silt-discharge channels were incorporated, that reservoir is predicted to "reach equilibrium with silts" by 2020.

The Mother River has never been easy to control.

National Parks of Ningxia

The moment we stepped off the plane in Yinchuan, we were struck by the changed climate: although the sun was about to set, the air was noticeably warmer and drier than it had been in eastern China. We were in Ningxia Autonomous Province, about 600 miles from the east coast of this vast nation and just south of Inner Mongolia (38°30'N; 106°20'E). Annual precipitation there is only about 12 inches. Signs were written in both Chinese characters and Arabic script.

Yinchuan city sits beside the Yellow River in the shadow of the Helan Mountains. We visited several national parks in the area including Sand Lake, known for its birdlife. A huge entrance arch in the shape of a crane greeted us, along with a bank of fee collection stations. One must pay the entry fee, then also buy a boat ride, the only way to really see the lake and sand dune island. We decided to sidestep the steep entry fees and enjoy the birds to be seen outside in nearby fish-rearing ponds, including Grey Herons, River Lapwings, Great Crested Grebes, Grey-Headed Lapwings, White Wagtails, Egrets, Black-Necked Stilts, and Hoopoe.

The same approach to infrastructure was repeated in other nearby parks: massive, gleaming entrance buildings with marble staircases and plazas, along with high entry fees plus extra charges for trams, boats, and films. It is an interesting approach to park management—landscape preservation coupled with costly ecotourism playgrounds.

The water at Sand Lake comes from the Yellow River. The abundance of surface water, there and in fish-rearing ponds that lined the highway, was a bit startling, since water use along the river is tightly controlled by the Yellow River Conservancy Commission. Most of the diked ponds were former wetlands converted to fish culture. China has a long history of inland aquaculture and ranks first in the world in commercial fish rearing. Freshwater aquaculture includes various species of native Carp plus exotic Tilapia from Africa.

The visitor center and museum at the Helan Mountains National Nature Reserve had two floors of museum displays and replica petroglyphs. We were happy to move outdoors and walk the canyon trail with the real rock art. There, hundreds of intricate petroglyphs are pecked into the stone, some of them dating back 6,000 years.

A small creek trickled along the bottom of the canyon. The park had informative interpretive signs, and small dots were painted near some of the artwork that, having faded, might otherwise be overlooked. "Chinglish," is what our guidebook called the language on signs, some of which bore humorous translations; still, they managed to communicate key messages:

> *"Please appreciate the sceneries and the historic sites as you want to be appreciated."*
> *"Well treat the environment and nature."*
> *"No mountaineering for hillside steep the stone loosening!"*

Kinder was interested in our reactions to the signs, wanting to know if we understood the messages. Here again, his English skills and connections with park interpreters may be able to help make the translations more accurate (while perhaps taking away some of the charm of the signs).

Even in that pristine setting, park managers could not resist remodeling the creek bed into ponds lined with stones. In some of these man-made pools, large Carp (goldfish) swam.

Toward sunset, we were back in the city of Yinchuan at the Yuehai Wetlands National Park. Our driver dropped us at the bottom of a huge staircase that led up to a stadium framed by Romanesque arches. Walking through an arch, we would not have been surprised to see gladiators battling in the arena below. Instead, players were kicking a ball across a soccer field at the edge of a large reservoir. The water came from a tributary of the Yellow River that runs through the city. While wedding couples were photographed amid the columns in the sunset glow, fishermen and birds enjoyed the lake. The man-made lake and its grandiose structures were built to celebrate the fiftieth anniversary of the Ningxia Autonomous Province's creation. One of the lake's justifications was to restore wetland acreage previously lost to farm drainage in that region.

The amount of total water allotted to each of the nine provinces through which the Yellow River passes is set by the national YRCC. In Ningxia, with 15 percent of China's irrigated farms, agriculture uses about 93 percent of the province's allotment, but that number is changing. To provide more water to cities and to coal-based industries, the government wants farm water use to drop by 22 percent. Construction of new factories and power plants can proceed only after irrigation canals serving farms are lined or repaired, to free up "new" water. By 2010 three such water-trading projects had remodeled 37 miles of canals and 105 miles of substreams in Ningxia, making 13 billion gallons of water a year available to new coal-burning power plants (Ivanova, 2011).

From Yinchuan, we traveled south, paralleling the Yellow River in a car powered by compressed natural gas. We were pleased by the skills of our hired driver, but had to stay calm in the backseat while his frequent passing turned the ride into a pulse-raising adventure—sometimes three cars occupied the full width of the two-lane roads. It seemed best not to watch. Along the river-fronting highway, we passed more channels diverting water to fish ponds, saw sheep and donkeys (on the east coast there were few grazing animals), and also spotted an aluminum factory and a magnesium factory on the riverbank, both busily pouring smoke into the air.

Driving, instead of taking an overnight train, was a last-minute change of plans that allowed us to see more of the countryside and to stay a night inside Shapatou National Nature Reserve, near the city of Zhongwei. About halfway there we stopped when our GPS told us we had reached 38°00′N. A tall river-viewing tower was under construction outside the village of Qingtongxia Zhan, though not yet ready for use.

Shapatou is called "the hat of the desert," because it perches at the very edge of the Tengger Desert, where advancing sand dunes are halted by the Yellow River (37°27′N). Silk Road travelers crossed this water barrier on leather rafts. The region was short on trees for wooden boats, but was rich in livestock, so they devised an ingenious solution to the problem: they soaked sheep hides in oil and brine to make them airtight, sewed the legs shut, and then inflated the skins like balloons. Lashed side by side, the hides kept wooden platforms afloat. In Shapatou National Nature Reserve, tourists ride on rafts supported by twelve to fourteen skins and navigated downstream in the fast-moving current by a paddler who sometimes teaches his passengers a traditional boating song, "The Yellow River Has 99 Bends."

The meeting of river and desert at Shapatou makes a dramatic natural setting, but there, as in the other national parks, moneymaking ecotourism activities abounded. We rode camels, the "ships of the desert," to the summit of dunes that loomed 300 feet above the river. That one-way ride deposited us near starting points for overnight camel-caravan camping trips; there were dune buggies to rent and race around on a marked course; zip-line cables spanned the river for sliding over and back; bungee jumping was available from a platform hanging out over the water; and sand-sliding was a quick way down the hill.

Seeking more natural experiences, we glissaded down the steep face of the dunes—a new experience for Kinder (this was also his first visit to the region)—and crossed a footbridge over the river to a marsh alive with birds. A section of the Great Wall was nearby, crumbling but still

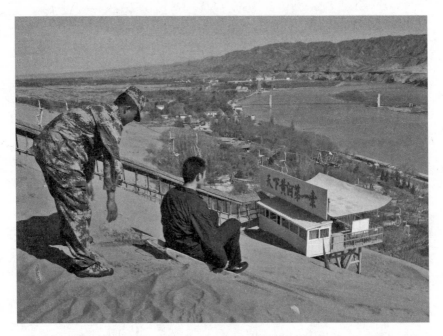

Figure 4.   The Yellow River stops advancing sand dunes in central China at the Shapatou National Nature Reserve, which offers ecotourist adventures.

impressive. Its effectiveness as a frontier barrier must have been enhanced by the nearby river. We were 600 miles west of the better-known sections of Great Wall we had seen outside Beijing.

Highly detailed museum exhibits in the park explained desert natural history and the work carried on by the Desert Research Institute to combat the expansion of deserts. Eighteen percent of China's land is threatened by desertification, an increase in unusable sand and gravel soils caused by overcultivation, overgrazing, and logging. In China, where global climate change is not a controversy that is avoided, the displays explained that climate warming is accelerating desertification problems. The onset of spring arrives 2.4 days earlier than it did in the 1980s, and China's glaciers have undergone a 21 percent reduction in the past 50 years. Since the institute pioneered a system of straw thatch squares to control blowing sand, the "Shapatou mode" has been applied to almost

400,000 acres, allowing vegetation cover to increase from less than 1 percent to over 42 percent in the most successful treatment areas. The effort began a half-century ago, motivated by a need to keep sand off the Bao-Lan Railway tracks that run from Inner Mongolia south through Gansu Province. Despite the successes of the Shapatou method of dune stabilization and the national tree-planting campaign we had seen along every highway, Chinese officials predict a 300-year effort to achieve stabilization.

A poetic greeting was displayed in the first room of the museum:

Here, the rolling Yellow River frames an "S" type basin,
becoming quiet tenderness . . .
Here, quiet and ancient houses stand with 100-meter high hill, sand, mountain,
*   river, park, symbiotic harmony, thousands of years living together . . .*
So, dear friends, forget your troubles,
*   hugging to the desert,*
*   exploring the scientific truth,*
*   feeling nature's purity;*
Maintain the natural balance with your heart.

We wandered into the Desert Research Institute, which is affiliated with the China Academy of Sciences. The compound seemed abandoned until we located a graduate student working by herself in a laboratory with flasks of water and soil. She told us she was analyzing relationships between soils and plant restoration. About forty faculty and students use the institute's facilities, but that day everyone else was in the field.

Despite the institute's efforts, desertification remains a major concern in western and central China. High winds on our second day in Shapatou gave us a small taste of the potential for sandstorms, which can be so severe in the spring that haze from China swept into the jet stream whitens the sky above our Sierra Nevada home.

Our rooms were inside the park, only 50 feet from the edge of the river, which we could see out the windows. Kinder said the hotel was "not bad for a national park," after showing us the trick for stopping the running toilet. (Toilets are not the subject of this book, but a tip for travelers to China: carry your own supply of paper. There may or may not be

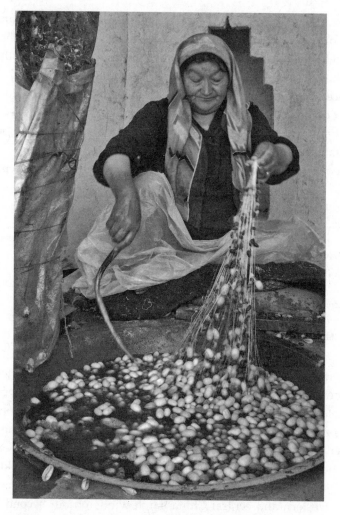

Figure 5. In Hotan, famous as a southern Silk Road oasis town, silk thread is gathered onto spools from silk moth cocoons.

any, even in new hotels with beautiful decorative paper dispensers). In the restaurant beside the hotel a chef showed us a live fish, caught that day in the Yellow River, that he then cooked for our dinner.

Hide boats and camels at this ancient river crossing were evidence of the Silk Road history that we would continually encounter from there across Asia. Camel caravans moved steadily between China, Europe, and India between the middle of the first millennium B.C. Activity along the Silk Road routes peaked during the Han Dynasty (200 B.C.–220 A.D.) and surged during the Tang Dynasty (618–907 A.D.) and again while the Mongols ruled northern China (1220–1368 A.D.). From China came silk, iron, steel, bronze objects, lacquerware, and hides. Europe and India bartered glass, gold and silver objects, incense, amber, pigments, ivory, and slaves. When the Ming Dynasty sealed itself off from the outside world, near the end of the fourteenth century, activity ceased.

Trade was also ongoing within the region itself. China required horses for defense against attack from nomads. The horses were, ironically, acquired by trading with horse-rearing nomads. During the Tang Dynasty, China paid the Uygurs 1 million bolts of silk for 100,000 horses each year.

According to his famous diary, in 1273 Marco Polo traveled east on the southern Silk Road, moving from oasis to oasis. The young man's father and uncle had come to the court of the fifth Mongolian ruler, Kublai Khan, in 1265. A year later, they carried a message to the pope requesting one hundred missionaries be sent to the khan's court. That never happened, but two Dominican monks, plus Marco, began a return trip. More interested in seeing Palestine, the monks left the group in Syria, but Marco reached Kashgar, then Yarkand and Khotan (today's Hotan). He dictated his famous descriptions of the following 18 years he spent in China to a fellow prisoner in 1298, while he was incarcerated during a war between Venice and Genoa.

The Taklimakan Desert was—and is—the most daunting obstacle to travelers moving across western China. The name Taklimakan, in the Uygur language, means "the place where travelers will never get out." At Lanzhou, northern and southern routes split to circle that desert, which is about one-third the size of California. The southern branch passed through Xining then skirted the edge of the great desert along a string of oasis towns. The 38th parallel traces that southern portion of the Taklimakan, so the second half of our journey across China would take us first to Lanzhou and Xining, before we leapfrogged by air over the vast distance to reach Hotan, Yarkand, and Kashgar.

Up the Yellow River
to Lanzhou's Green Camel Bell

The city of Lanzhou was an important Silk Road hub that became a major industrial center for modern China. With 3.2 million people, Lanzhou is the first major city on the Yellow River after it flows out of the mountains. The city is south of the 38th parallel, but just as deserts and mountain ranges once funneled Silk Road caravans, major highways now follow the Yellow River south before continuing westward.

Blowing sand and dust turned the sky white as we approached Lanzhou. When airborne, sand does not appear light brown, but fills the sky with small haze particles. Major sandstorms were then occurring across northwest China, though local conditions were not so severe.

Lanzhou hugs the banks of the Yellow River for 24 miles and is confined by hills on either side of the river gorge. We checked into a Super 8 hotel, the American chain, with landscape photographs of Arizona and

Utah on the lobby walls and a Route 66 sign in our room. A short walk brought us to the city's central square, where large groups of men and women were doing a popular Mongolian dance as we came out of a supermarket. As the sun went down, a garbage truck drove by blaring music like an ice cream van in the States. The song, on that warm May night, was "Jingle Bells."

Ran Liping met us with bright eyes and an infectious smile; setting other projects aside, she devoted the next two days to us. Liping is a project coordinator with the NGO Green Camel Bell (GCB), whose mission is to protect the environment of western China, seeking "green mountains, clear water, blue sky, man and nature in harmony." The organization's unusual name was chosen, she told us, because "green is our dream—where we hope to go will be green; camels are well adapted to their environment, as are the GCB staff and volunteers, but can get lost in the vast desert, so ringing bells help people know where we are, and they bring luck."

We also met the NGO's founder, Zhao Zhong, who received a *Time* Magazine "Hero of the Environment" award in 2009. GCB's full-time staff of six people work with two interns, two part-time staffers, and about thirty volunteers on a wide variety of projects including grass-brick construction, sustainable agriculture, environmental education, and community development. The organization has an impressive list of international supporters from the United States and Europe, including, since 2006, Pacific Environment, headquartered in San Francisco. On one office wall were logos for twenty-six organizations, including the World Environmental Fund, International Rivers, the Global Greengrants Fund, the World Keeper Alliance, and the Ford Foundation.

Liping began working for GCB as one of its first staffers after she completed her degree in environmental science in 2008. "When I graduated, I thought I should do good things for the environment, so I chose this job," Liping said. "I think the youth should take responsibility for the environment," she added, using words almost identical to those spoken by the younger activist JungRok in South Korea.

GCB works mainly in Gansu Province, which extends north and south between mountain ranges and desert. The organization became heavily involved in reconstruction after the 2008 earthquake in Sichuan Province (southwest of Gansu), where a magnitude 8.0 quake killed at least 68,000 people. In Wenchuan village, near the epicenter, GCB has been supporting reconstruction with green techniques, such as using straw-bale bricks instead of cement and providing ecotoilets that need no water (the whole effort helped by Friends of Nature).

In the center of Gansu Province, GCB persuaded farmers who traditionally grow wheat, corn, and potatoes to add sunflowers as a crop, because the growing plants add nutrients to the region's alkaline soil and need little water. They help farmers to form cooperatives and have provided some farm machinery, again with help from international groups. Farm family annual incomes doubled in 2009, rising by $30.

Most of the region's economy is related to the Yellow River in some way. University students are trained by GCB to be environmental educators in primary and middle schools. Their program theme in 2011 was "Yellow River history, culture, and pollution." Teams of "river protectors" are also trained to monitor the water's oxygen demand level, an indicator of pollution levels. They post the data they collect to the national Environmental Protection Bureau's water pollution map. "We are the third audit," Liping said, "the independent party that checks these things, along with the local and national governments." Though GCB has three hundred factories on an office blacklist, only a few of those are on the national map. The *Time* article "Heroes of the Environment, 2009" pointed out that "multi-nationals and their offshoots have taken note [of GCB's audits]. After a brewery it part-owned in Tianshui ended up on the map, beer giant Carlsberg installed wastewater-treatment facilities. State-owned firms that are found polluting usually feign indifference, but at least they know they are being watched, a new and uncomfortable experience" (Ramzay, 2009).

At the time of our visit to Lanzhou, only 30 percent of the city's sewage was treated before going into the river, but a new sewage treatment plant was under construction. Once it begins cleaning the city's wastes in 2012, about 90 percent will be treated. Before 2006, *all* of the city's sewage went straight into the Yellow River, along with chemical, oil, and other industrial pollutants.

"Because we are on the upper watershed of the Yellow River, we must be responsible about not polluting the river for domestic users," Liping said. She was especially proud of her successful project to improve drinking-water quality in the small village of Liangjiawan (population 1,500), upriver from downtown Lanzhou. After a low flow-through dam was built across the Yellow River in 2007 to generate hydroelectricity, water began eddying back upstream. The nearby village had always taken its drinking water upriver from the point where its untreated wastes entered the river. Now the village's intake pump began sucking water contaminated with sewage. Liping and Green Camel Bell helped organize the village leaders to seek changes and secured water purifiers for each village home, four purifying machines for the village school, and a new water pipe system and purifying chemicals to serve the community.

Why would a dam be designed so poorly? China's laws do require an environmental review, but the flawed process did not identify the problem. Because the dam generates electricity with flowing water—a clean alternative to coal—the company secured carbon trade credits, and the impact on local people was not addressed.

We went upriver with Liping that afternoon to see another village with a slightly different drinking-water problem. It was nice to be outside under blue skies with temperatures, Liping said, as warm as summer. Although Lanzhou has long been notorious for severe air pollution, winter is when the problem worsens, once everyone begins heating with coal.

Another small hydroelectric dam had recently been completed across the Yellow River, downstream from Cxian Shui, a small village of narrow dirt lanes that hugs the riverbank. This village depended on clean water from a community well that, two years previously, was drilled deeper to get below a layer of saltwater. Liping approached several ladies in the village to ask about their water. They seemed reluctant at first to talk, but Liping's charm and sincerity won them over. We heard that new pipes had broken and were never repaired, so now they were hauling drinking and cooking water in buckets from the Yellow River, which had to be boiled. Some people had plastic water pipes running above ground to the river and fed by submersible pumps. Another option was a 10-minute walk to the well lugging heavy buckets. Once they began describing their plight, the women's frustration became obvious.

The young baby we met and all the people in that village deserve better. Safe and clean drinking water and sanitation are basic human rights, affirmed in a United Nations resolution in 2010. Liping's next step will be to contact village leaders to confirm the stories she heard from the residents and begin working toward a solution.

Green Camel Bell is a pioneering group in China, helping to effect change in attitudes regarding the environment. Liping told us that the staffers are all unmarried and completely dedicated to their work, but that some find it hard in the present Chinese culture to explain the value of their work to friends and family.

The first Chinese NGO was founded in 1994. Friends of Nature focused on environmental education at first, until the cautious central government made it clear that its role could broaden. By 2006 more than 350,000 NGOs were officially registered with the Ministry of Civil Affairs. With rapid economic growth driving China's society, enforcing compliance with national environmental regulations had become difficult. The NGOs serve as that useful "third audit" described by Liping.

On our second day in Lanzhou, our guide Kinder departed for Beijing to meet another group. With Liping, we went to see 60-foot-tall waterwheels, ingeniously powered by the flowing current of the Yellow River to lift water high above the riverbanks. Similar mills have served this city, local farms, and grinding mills ever since the Ming Dynasty (1368–1644 A.D.). In 1952 Lanzhou, "the waterwheel city," still had 252 of the big wheels lining both banks of the river. Some of those still operating are hundreds of years old.

Massive logs in the waterwheel frames were evidence big trees once grew on nearby forested hillsides that today are mostly treeless. Beside the highway coming into Lanzhou, terraces were planted with small conifers, part of the national tree-planting campaign. Japanese groups have even come to Lanzhou to plant some of those trees, because the spring dust storms blow their direction.

Liping arranged a meeting at a teahouse with one of her university professors, Zhang Songlin, who teaches geography and environmental science at Northwest Normal University. She translated as the professor told us about sustainable development issues in China. "All countries need to have sustainable development," he said, "but many of our local governments focus on [fostering] higher GDP." Throughout China, whenever the subject was economic growth and development, the speakers used the shorthand GDP for gross domestic product.

"Officials, while serving their terms in office, need good results so they can go out at a high rank. People need jobs. Young people need an income to buy a house, get married, and become self-sufficient. But the old Chinese culture respected nature. People and nature were thought of as one. In old times, the belief was that if you harm the environment, you would be punished. Now, increased GDP is the belief. We don't trust each other enough. We can't trust food safety. Air quality is improving little by little, and also water quality, but I am not very hopeful about the future." Despite that pessimism, Professor Songlin agreed with us that Green Camel Bell's work was a reason for hope.

That night, Liping escorted us to the train station as we were heading west to the city of Xining. She would undertake her next task in Wuwei, a city right on the 38th parallel, where she would begin still another water-quality investigation.

Qinghai—Blue Lake of the Tibetan Plateau

Xining is the home of the Qinghai (pronounced "ching-high") Salt Lake Institute. We had seen a news story in the Chinese press in 2008 headlined "Massive Program Launched to Save Qinghai Lake" (Xinhua News Agency, 2008). That parallel to the historic efforts to save Mono Lake, plus the fact that China's largest saline lake occupies a watershed intersecting the 38th parallel, made Qinghai Lake, east of Xining, a must-see destination.

A researcher at Mono Lake wrote to introduce us to Professor Fafu Li from the institute, who was our gracious host for several days. Fafu arranged a meeting with institute scientists and graduate students interested in hearing about Mono Lake and our 38th parallel project. He had worked for 19 years at the institute.

One of the faculty members at our meeting was Dr. Junqing Yu, the

Figure 6.   Qinghai Lake's alkaline water generates suds when waves lap the shore.

author of a book (written in English) about Qinghai Lake. He particularly wanted to compare the chemistry of Mono with the vast inland sea
of central China. Qinghai Lake is saline, though less salty than Mono
Lake or the ocean (at 14 grams per liter dissolved salts). Its water is quite
alkaline, though, with a pH of 9.2 (Mono Lake water is pH 10, similar to
mild detergent). Qinghai is also supersaturated with dissolved aragonite
and calcite, the two common crystal forms of calcium carbonate and the
main ingredients in Mono Lake's tufa towers.

Lakes like Qinghai and Mono that occupy closed basins with no out-
let gradually accumulate salts dissolved from the watershed by tributary
streams. The Earth's oceans are salty for the same reason. Qinghai Lake's
watershed valley, drained by six rivers, covers more than 11,400 square
miles. Most of its water arrives via the Buh River, which extends northwest almost 200 miles to the crest of the Datong Mountains, just north
of the 38th parallel. The Qinghai Nanshan Mountains, south of the lake,
approach the shore more closely. As we saw during two days of exploring, the Nanshan provide a scenic snow-covered backdrop, something

like the mountain setting at Mono Lake. Qinghai's water surface covers 1,700 square miles—a huge inland sea vastly bigger than Mono Lake, more comparable in size to Utah's Great Salt Lake. As the only large water body in central China, it stands out on maps.

Fafu and a driver took us to the lake, pleased, it seemed, to be escaping from the office on an overnight road trip. From Xining we crossed a pass at 11,250 feet above sea level and left the Yellow River watershed behind. The surface of Qinghai Lake is 10,154 feet above sea level.

It took most of the day to travel around the gigantic lake to its western shore, 220 miles from Xining. Grazing animals abound in the grasslands along the lake, and we got our first close look at yaks (*Bos grunniens*)—thousands of them and even more sheep. The diet of the local Tibetans is almost entirely meat, especially mutton.

Buddhist prayer flagging was strung in many places around the basin. The Mongolian name for the lake, Kokonor, means "blue lake." Both Mongolians and Tibetans consider it a sacred place. Here on the shore of the vast sea in 1578, a Mongolian leader conferred the title *dalai lama* for the first time, to a spiritual leader named Sonam Gyatso. *Dalai* is Mongolian for "ocean," while *lama* is a Tibetan word for "priest."

The main tourist destination at Qinghai Lake is Niao Dao National Nature Reserve, also called simply Bird Island, on the west shore. Each year, about 100,000 birds encompassing 189 species migrate through or nest there. Bar-Headed Geese (*Anser indicus*) arrive in the spring after crossing the Himalayas from India at elevations over 25,000 feet. Unlike the cranes of South Korea and the Yellow River Delta, which winter on the 38th parallel but nest far to the north in the spring, these waterfowl moved up from the south, where they migrated to avoid the harsh winters on this, the Qinghai-Tibetan Plateau. Another pattern would be evident once we reached wetlands and lakes on the 38th parallel in Europe and North America, where migratory birds stop only long enough to rest and refuel while traveling each spring and summer between northern nest sites and "winter avoidance" ranges in the Southern Hemisphere. Migrations are ultimately driven by the seasonal freezing and thawing of water, which forces plants and animals to adapt to protect body cells that are, essentially, complex concoctions of water.

After the long car trip, we were eager to walk, and so enjoyed the 200-yard stroll through an underground tunnel, wide and well lit with photographs of wildlife and scenery, that leads to a viewing blind where nesting geese were just a few yards away on the other side of glass windows. Thanks to the lengthy approach that keeps people out of sight, the birds were unaware that dozens of humans were within a few yards, watching their intricate courtship displays and flights. Here, the Chinese inclination for grand structures had produced the most spectacular bird blind we had ever seen.

An electric shuttle bus dropped us at another observation point above a nesting colony of Common Cormorants (*Phalacrocorax carbo*) and Black-Headed Gulls (*Larus ridibunus*). A trail led down to the edge of the lake. The water tasted slightly bitter, and we smiled at the familiar sight of suds forming where waves agitated the water on the shore. The lake supports diatoms and various kinds of algae that feed just one kind of fish, the endemic Scale-less or Naked Carp (*Osteichthyes cypriniformes*).

The Carp never showed themselves. They have been protected from fishing for almost 20 years, yet at the hotel restaurant that night our waitress asked Fafu if we wanted to order the fish! He declined and told us that it would be "very illegal"; he could lose his job if he had said yes. It seemed a fishy situation that the protected Carp was available there on a menu that otherwise featured meat and potatoes.

Grassland surrounds Qinghai Lake in an almost treeless landscape where a limited water supply makes crop growing difficult. Sheep, goats, and yaks graze the grasslands, competing for forage with diminishing numbers of wild Przewalski's Gazelles (*Procapra przewalskii*), an endangered species. Much of the grassland close to the lake has been converted to rapeseed (*Brassica napsus*) farming. That requires irrigation that has contributed to the decline of the lake, which lost 10 feet of its depth and shrank by about 150 square miles between 1959 and 2006. Publicity photographs for Qinghai Lake always feature the bright-yellow rapeseed flowers that bloom in the summer. The plant is valued for making canola oil, plus by-products of oil production that include animal fodder and "oil-cake" fertilizer. Fields were brown and stubble was burning when we were there; it was still early spring at that altitude.

The government's 10-year plan to "save Qinghai Lake" includes actions across the watershed to maintain the lake level and improve ecological conditions. Some irrigated farm fields have been taken out of production, sheep numbers are to be reduced by about 1 million animals, and 2 million acres of irrigated pasture are to revert to grassland. "Ecological migration" of the region's nomadic herders will relocate four thousand people.

To reduce livestock numbers, individual, fenced grazing parcels have been allotted to the traditionally nomadic people. In several places along the southwestern and northern shores, we saw new houses provided by the government. The nomads are still allowed to roam with their herds to higher ground during the summer, but total numbers of animals are controlled. Additional justifications, says the government, for the controversial forced housing change are clean drinking water, improved housing conditions, and better diets.

The government has also been poisoning rodents, which officials say degrade grasslands and compete for forage with livestock. This program is also controversial. Some researchers consider the targeted Plateau Pika (*Ochotona curzoniae*) a "keystone species" in the grassland ecosystem of the Tibetan Plateau, providing food for predators and burrows for birds seeking breeding shelters. The only pika we saw were mounted specimens in a museum.

On the north shore, Kangtsa is the largest town in the lake basin. It is an agricultural center and the capital city of a county with the same name. Nearby, a railroad line that connects the area to Lhasa was being converted for electric locomotives. New electric lines were running everywhere, some from solar and wind energy–generating plants. The high plateau must have looked very different just 20 years ago. The region illustrates a dilemma facing central and western China, where public services are improving, but at a cost to traditional ways of life.

After circling the north side of the lake on a very rough road, we stopped at Qinghai Lake National Park. There had once been a grand plan here for lake cruises. Docks and other infrastructure were built, but the government decided that boat engines were too polluting and halted the project. In fact, no boats of any kind are allowed. Although the potholed access road and abandoned facilities gave the park a ghost-town feeling, there was a bus group of tourists there that day, and a crew was stabilizing sand dunes using the straw squares developed at Shapatou. The elevation, clean air, and endless vistas reminded us of home.

Before leaving the Qinghai Lake area, we crossed a 12,500-foot-high pass into a separate basin and explored hypersaline Chaka Lake. Only bacteria live in its concentrated brine, as in the Dead Sea of Israel. Saltworks have been here for centuries. Fafu told us that Chaka could provide enough domestic salt for the entire world for 500 years! Large barges moved through a channel past evaporation ponds toward the middle of the lake. We walked, far from the open water, along a jetty lined by a layer of salt a foot thick.

While we lunched at a small noodle cafe that mostly served locals, the waitress suddenly told us to hurry and finish because a serious sand-storm was blowing in. We barely beat it back to Xining.

Under skies turned white by blowing sand, we flew above the Taklima-kan Desert to Urumqi, in the Xinjiang Uygur Autonomous Region.

Hotan to Kashgar
on the Southern Silk Road

The tourist map for Xinjiang states that Urumqi, the transportation hub in northwest China, is the city farthest from any ocean in the world, over 2,000 miles from Beijing and the east coast of China. Our next goal was actually Hotan, one of the oasis towns along the southern Silk Road to Kashgar. Resolving some confusion arising from the Chinese spelling of that city as "Hetian" on airport signs, we made our connection and were happy to see our Uygur guide, Abdul Wayit, and a friend from home, Rick Kattelmann, waiting for us at the Hotan airport. Rick is a retired hydrologist, now a serious photographer, who had traveled here before and recommended Abdul's guiding services. Abdul, 23 years old, handsome, and ambitious, had learned English and become a guide as a way to leave the family farm. He was eager to share his Uygur culture and became an invaluable window into the soul of his people.

Figure 7.   At the edge of the Hotan oasis, men plant peach trees to keep the Taklimakan Desert at bay.

Hotan is famous as an oasis on the Silk Road and especially for its jade and silk. The sprawling settlement is an island of green in the Taklimakan Desert, with orchards of fruits and nuts and fields of wheat going on for miles. At the eastern edge of town, men were planting and drip-irrigating peach trees and tamarisk as part of the "green wall" that holds back the shifting dunes.

Hotan exists because of its water—the White Jade (Yurungkax) and Black Jade (Karakash) Rivers, which flow from the Kunlun Mountains. After the two merge north of the city, they become the Hotan River, which continues north and soon disappears beneath the desert sand, except during the wettest seasons. Hotan traded in jade with Mesopotamia from the third millennium onward. A new discovery in 2001 in the riverbed gravels started a modern-day jade rush, a frenzied free-for-all in which miners have torn up the riverbed using everything from shovels to huge excavators. People sleep among the tailing piles and work with lights at night, looking for the shine of the precious stones. In an informal jade market near the river (a parking-lot gathering), men haggled over stones. One jade buyer showed us a golf-ball-sized piece that, he said, was worth 5,000 yuan (about $1,600). He glanced at a small rock we found and tossed the worthless river stone contemptuously on the ground.

The jade found here is rarely green. Most of it is "mutton-fat jade"—opaque, creamy white, with tan or orange tinges. Abdul told us that many people in Hotan had become wealthy by prospecting, as the market price per ounce is as much as forty times higher than that for gold. "Eighty percent of the people in Hotan now have cars," he said, attributing that to jade trading. Local Uygurs sell their finds to Chinese middlemen who serve the demand from growing numbers of wealthy Chinese. In Chinese culture, jade is the most valuable of precious stones, symbolizing perfection, constancy, and immortality.

Those ethereal qualities contrast with the toil of thousands of miners and hundreds of large earthmovers that we saw turning over the riverbed. The channel looked like the aftermath of the California Gold Rush, where mounds of dredged gravel are still seen, 150 years later, heaped along the banks of foothill rivers.

When melting snow sends high flows down toward Hotan, the torn-up channel can no longer contain the water, so floods have become a threat (as they did in California by 1860). In 2007 the Chinese government issued a notice to stop the heavy-equipment work, but prospectors ignored the order, and local officials did not enforce it. We saw plenty of excavators at work, along with small groups of people wielding shovels.

Silk, rather than jade, has been Hotan's most famous and enduring trade item for centuries. A factory in Hotan still does the whole process by hand, extracting the thread from the cocoons of silk moths, then spinning, dyeing, and weaving fabrics for garments and tapestries.

Farming has been the traditional support for most of the population of Hotan, which is renowned for its almonds and walnuts, plus fruit and vegetable staples. Thanks to irrigation, the greenery of the oasis city was more extensive than we expected, though it was just a short drive from the center of town to the edge of the desert.

Walking across dunes gave us a feel for the abrupt change. A half-mile into the desert, a tomb and small mosque marked the burial site of

Imam Asil, who died about 900 years ago during a war between Uygur Muslims and Tibetan Buddhists that went on for hundreds of years. Pitchers with water for washing before prayer were being filled from a well. A bucket is lowered and raised dozens of times a day. We took turns at the crank, confirming something we already appreciated: water is heavy—eight pounds per gallon, hauled up 20 feet at a time.

That night in Hotan, Abdul asked if we wanted to "eat the food Marco Polo took back to Italy." The spaghetti was delicious, though Marco Polo did not eat his with tomato sauce in the thirteenth century. The legend of Polo introducing pasta to Italy is a popular story in that region, but Italians actually had been making pasta for centuries before Polo came to China.

Driving west from Hotan, we stopped in the village of Kargilik to explore a colorful weekly market. Underneath red tents, the colors of silk clothing and textiles glowed with fiery brilliance. This oasis was the junction point with a route to Ladakh over passes through the Himalayas to the south.

We continued westward and northward toward Yarkand and Kashgar. Even with an air-conditioned van and a good driver, the 350-mile, often bumpy trip felt long. What must it have been like to do this same trip on a slow-moving camel? A railroad is set to open soon along the route, and a new highway was under construction. The evolution of the Silk Road continues.

It was a relief to break up the trip with a night in Yarkand, a small town by Chinese standards, with only 560,000 people (38°17'N), surrounded by the largest oasis in Xinjiang Province. Water comes down the Tarim River out of the Karakorum Mountains, then flows northeast year-round along the northern edge of the desert. On a walk through the oldest section of the city, we asked Abdul about a sign painted on a mud wall. "We only have one Earth, so practice family planning," he translated, adding that the government had placed the sign there.

"We agree with that message very much," we told him. "What do you think?"

"I think it is good," Abdul said, then grinned and added, "but I still want five children." Large families are a way for Uygurs to fight back against what they perceive as an invasion by the Han Chinese, who keep moving to the western province. Although the government limits rural Uygur families to three children, it is apparently possible to circumvent the limit if babies are born in different localities.

Northwest of Yarkand, the desert suddenly turned green again when we passed a reservoir outside Yengisar, "the town of knives." Abdul asked if we wanted to stop and shop for a knife, but of course there was no way to bring one home in our carry-on luggage. He told us that Uygur men used to always carry knives on their right hips. "Our knife is our manhood," he explained. But after violent demonstrations in recent years, men are no longer allowed to openly carry a knife. "If I wore one now, soon I would belong to the police," Abdul said.

The Uygur region seems a whole different country than the rest of China. For much of human history, it was. Veiled women mix in the business districts with Han Chinese women in miniskirts and high heels, but the two cultures segregate into separate residential areas of the cities. After China took over in 1948, the province was named Xinjiang, meaning "new frontier." The Han living in China's crowded east are encouraged to move west. The ensuing influx has generated tension that has erupted in violence in Urumqi and Kashgar. Just a year before our trip, our friend Rick had toured the same region and seen everywhere a heavy presence of military and police. Happily, things had calmed down since then. (But a year after our visit, violence flared again, and the police shot several demonstrators.)

Along the highway, a truck was parked in our lane while the driver negotiated with a sheepherder for some of his animals bunched nearby. What sheep found to eat in that desert was a mystery; it appeared nearly barren (in part because there were so many sheep and goats nibbling whatever grew). The purchased sheep were likely going to the Sunday animal market in Kashgar.

Kashgar is famous as a former Silk Road trading center and current crossroads bridging China and Central Asia. Pomegranates and apricots are the specialty crops, grown near the city with water from the Kashgar River. The city's namesake, this stream descends from the Karakorum and Kunlun mountain ranges. Marco Polo wrote in the thirteenth cen-

Figure 8. The Sunday animal market in Kashgar, where sheep, goats, cows, donkeys, and racing horses are for sale. Note the money held by the man making an offer to the rider.

tury that the "inhabitants have very fine orchards and vineyards and flourishing estates. Cotton grows here in plenty, besides flax and hemp. The soil is fruitful and productive of all the means of life. This country is the starting-point from which many merchants set out to market their wares all over the world" (Polo, 1982 [1958], 65).

Outside Kashgar, we visited Abdul's family farm, where he was born and raised, and met his mother, father, and wife. It was a great honor to be invited into a Uygur home. Knowing they were part of a conservative culture, Janet asked if she should cover her head with a scarf while in their house. "No," Abdul answered, "they understand you are not from here."

The farm grows corn, wheat, and apricots (the latter are often eaten green), and the family raises a few cows, goats, and chickens. There's also a friendly cat. Abdul's wife cut and stretched noodles, her flashing fingers working the dough before she cooked the pasta. We gave his mother a Mono Lake shopping bag, and she seemed thrilled after she saw how it stuffed into a small sack.

The famed markets of Kashgar are incredibly colorful, with dramatically dyed fabrics, luscious fruits, and shiny copper pots. Women go to market in exquisite dresses that, Janet said, she would consider wearing only to a formal wedding. Livestock are brought each Sunday to a special market where sheep, donkeys, cows, goats, and even cats and dogs have been sold or traded for centuries. There was an area to test-ride sale horses at top speed.

On the way to the market, we drove by Kashgar's street market for mobile phones. In the twenty-first century, people buy and barter for items that have been manufactured for thousands of years, while carrying on phone conservations.

Many buildings in Kashgar's old town were being torn down and rebuilt with fired bricks and reinforcing steel instead of mud bricks. The government says the changes are necessary to protect against earthquakes, but Abdul believes the old city withstood many quakes with no problems. Forced renovation contributes to a perception that the Han Chinese are undermining the Uygur culture. The government insists that it respects minority cultures and pledges to keep most of the old town intact as a tourist attraction—with an admission charge.

Walking along Kashgar's streets, as in every city in China, can be a terrifying adventure. Chinese drivers do not slow down for pedestrians. One learns to follow the herd at crossings, to wait at the edge, and again on the line in the middle of the highway, as cars speed by too close for comfort, then proceed when it is safe. Traffic was less hectic in the old town, where we saw abalone shells for sale, though we were thousands of miles from the ocean. The shiny material is used to ornament items like knife handles. After watching a man skillfully turning wood on a lathe, we bought several small items—keepsakes from that oasis trading center on the twenty-first-century Silk Road.

The Edge of China and Turkmenistan

The Karakorum Highway (locals call it the KKH) connects China to Pakistan through the westernmost of the highway passes across the Himalayas. We were eager to compare the high country along China's western border to our home in the Sierra Nevada. A challenging barrier for Silk Road traders, the mountains are also the source of the melted snow and glacial ice that descends into the Taklimakan Desert basin to create the oases that made Silk Road travel possible.

A cutoff to the north from the main highway led to Oytagh Glacier, at the head of a scenic pastoral valley. *Oytagh* means "mountain basin." Backed by towering mountains, the valley community was raising crops along the river, while livestock grazed on the new spring grass. Yaks, camels, and Kyrgyz yurts, hats, and clothing appeared here, close to the border with Kyrgyzstan. It felt good to breathe clear mountain air. The river was muddy with water because the glacier, hanging below peaks towering 21,000 feet, was melting fast.

The Oytagh Glacier viewpoint is 9,533 feet above sea level. Just as we reached the viewpoint, an avalanche dropped from the cliff west of the distant glacier. To the left, a waterfall plummeted as straight as the upper Yosemite Falls, for hundreds of feet. Or farther, perhaps. Scale was hard to judge in this grand landscape.

The view from the end of the trail was once straight down onto the lip of the glacier, but the ice had retreated drastically and now was at least a half-mile away, leaving behind only frozen humps covered with black silt. Glacial retreat there was even more dramatic and obvious than in the Sierra Nevada. Some of the clearest signs of atmospheric warming can be seen at the world's higher elevations. Just as the glacial melting in Tibet is a threat to the Yellow River, rapid glacier melting—at a rate of 7 percent a year—in the mountains of northwestern China creates a water-supply concern for people downhill, as glaciers have been the dependable water source for rivers that serve the region's farms and cities.

Although we had planned to sleep that night in a yurt, we learned they were available only to Chinese tourists, not foreigners. Water was shut off inside the park, including the restroom buildings serving the yurts and even in the hotel built for foreigners, because of damage from frozen pipes that winter. Big jugs of water were provided for toilet flushing. The restaurant next door was also not quite ready for business. The staff offered to kill a sheep and barbecue kebobs, but we were too hungry to wait that long, so we made do with our own supplies. A thermos of hot water was left outside the door in the morning so we could make instant coffee.

Next morning, in the valley village, children gathered in their schoolyard to salute the flag while a recorded anthem played. A local bus was taking on passengers, including two live sheep being settled into the back luggage compartment, where suitcases are carried on urban routes.

The Karakorum Highway continued its climb up toward the Pamir Mountains. A herd of Bactrian (two-humped) camels grazed in a wash, and after photographing them, we asked Abdul if they were running wild. He said, "There are no wild things in China." (Happily, that is an

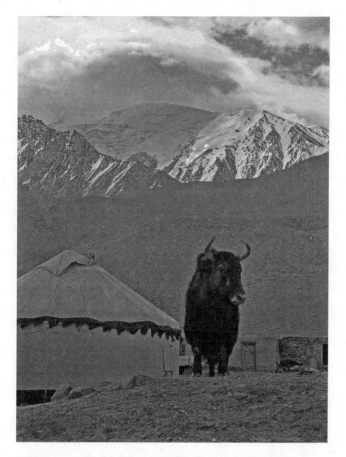

Figure 9. In the mountain valleys of western China, Kyrgyz people live in yurts and raise yaks.

overstatement; we had seen lots of wild birds, though few wild mammals.) The road also passed construction sites in the Gez River canyon for a hydroelectric energy–generating plant and, at the canyon's head, a new dam. An 800-foot drop from the dam would turn turbines to generate 200 megawatts of electricity. Most of the construction workers had been trained while working on the Three Gorges Dam on the Yangtze River.

Sand Dune Lake (not to be confused with Sand Lake near Yinchuan) will be elevated by the dam. We stopped to photograph yaks and yurts and immediately attracted men selling malachite and jade jewelry. There, at 10,770 feet, the road turned directly south toward Karakul Lake. In that remote location surrounded by mountain peaks, Abdul's mobile phone rang. He seemed surprised that we were amazed by the phone reception; China has exceptional coverage. The call was from our Chinese guide, Kinder, checking to see how we were progressing.

Travel guidebooks name Karakul Lake, 11,999 feet above sea level, the most beautiful lake in China. Clouds obscured the nearby mountain peaks that afternoon, but when we returned on the following day, there was a clear view of snow-covered Muztagh Ata, the 24,757-foot mountain that towers over the south end of the lake and sends water down the Yarkand River to serve oases in the desert. To the east, clouds continued to hover around two peaks in the Kongur Mountains, both higher than 25,000 feet. Kyrgyz nomads live near the lake and raise sheep, goats, horses, donkeys, yaks, and camels.

Janet went for another camel ride along the lakeshore. Those amazing creatures can handle everything from hot desert sands to cold, rocky mountain trails. Seeing them was a reminder that here, high on the Pamir Plateau, we were still on the Silk Road.

We drove up to 13,374-foot Tashkurgan Pass and passed the "Welcome to Tashkurgan Tajik Autonomous County" sign, written in Chinese and English. Here, at 74°55′E longitude, was our point farthest west in China, 44 degrees of longitude from the Yellow River Delta.

The highway plummeted almost 3,000 feet to an intersection where trucks were turning west toward the international border with Tajikistan, only seven miles away. We continued on the KKH, watching our GPS for exactly 38°00′N, and stopped there for photos in front of a small mosque not far from Takiman village.

About 70 miles south, the highway would crest again at Kunjerab Pass (15,397 feet) and the border with Pakistan. We stopped, instead, at Tashkurgan, just as Chinese caravans had done for centuries to transfer their wares to Central Asian traders (37°47′N). Beyond there, branches of the old Silk Road led south to Kashmir and India. The westward route continued through Kabul (Afghanistan), then south of the Caspian Sea via Tehran (in Persia, now Iran), before turning back toward the 38th parallel and the Mediterranean ports in Turkey.

Tashkurgan was our final destination in China. Fortress ruins on the hill above the city date to the eighth century, 500 years before Marco Polo passed through. The culture of neighboring Tajikistan dominates the Tashkurgan population, so traditional dress was again different, with women wearing pillbox hats covered by scarfs.

Crossing China had been fascinating, infuriating, mysterious, and wonderful. We took away memories of vast deserts punctuated by oases fed by snowmelt from amazing mountains, by rivers that refuse to be tamed, and by mostly wild parklands that persist along with crowded cities and the world's largest population. While China is still officially a Communist nation, capitalist enterprises burgeon across the country in the twenty-first century, and the central government does not have total control. It was a pleasant surprise to learn that NGO groups like Green Camel Bell are welcomed by the national government because they implement and help enforce environmental laws in distant provinces—a sign that China is not simply ignoring the unwelcome side-effects of rapid development.

Returning to Urumqi to fly west on the 38th parallel, we passed above desperately poor Tajikistan, war-torn northern Afghanistan, and a finger of southern Uzbekistan to reach Ashgabat, the capital of Turkmenistan.

When the GPS screen showed we had reached the 61° E longitude in Turkmenistan, we stopped the car. Back home, at Mono Lake, the 38° latitude line intersects the 119° W longitude; since 61° + 119° = 180°, we were halfway around the globe in Turkmenistan, where local clocks were 12 hours different from California.

Earlier in Ashgabat, the capital and largest city (37°58'N), we had been met by our next local guide, Berkeli Atayev, tall, dark, and welcoming. Berkeli had visited Great Salt Lake seven years earlier with a group of Turkmenistan tourism officials. This day, he drove us straight east for five hours toward Merv, the most influential ancient city they never taught us about in world history class. After weeks in China, the most immediate impression of Turkmenistan was how few people there seemed to be; the national population comes to just slightly more than 5 million.

Halfway to Merv we reached that halfway-round-the-world longitude point, beside a reservoir that stored Karakum Canal water. Completed in 1954, when Turkmenistan was part of the Soviet Union, the canal diverts 50 percent of the Amu Darya River, a major tributary of the Aral Sea. The canal supplies 90 percent of the nation's water and moves it 540 miles to Ashgabat and to farmland along the route, making it the second-longest aqueduct in the world. (San Diego reaches 444 miles into the Central Valley via the California State Water Project; the Los Angeles Aqueduct extends 338 miles into the Mono Lake Basin. The city of Sirt, in Libya, moves water the farthest—745 miles from sub-Saharan aquifers to the north coast [Shea, 2011]). With the water, Turkmenistan grows cotton, a crop farmed there for 2,000 years, as well as fruits, vegetables, rice, and wheat on the tiny portion of the country (4 percent) that is arable. Most of the nation is covered by the Karakum Desert.

Because of diversions from the river tributaries to the Aral Sea by Turkmenistan and other nations, the fourth-largest lake in the world lost 90 percent of its volume by 1970. Its water level dropped more than 60 feet, causing the sea to split into two small lakes. Salinity increased to lethal levels for fish, and a vast acreage of toxic salt-covered land was exposed to the wind. Downwind in Uzbekistan and Kazakhstan, respiratory diseases and throat cancer rates soared. The Aral Sea's problems originate in other nations, and so far, little has been done to effectively address one of the world's greatest environmental tragedies, though a dam was funded by the World Bank's International Fund for Saving the Aral Sea to wall off the remnant in Kazakhstan, which has improved conditions there, on a small scale, for fish, birds, and residents. In 2010 Turkmenistan hosted a meeting in Ashgabat with Uzbekistan, Kazakhstan, Kyrgyzstan, and Tajikistan—the four Central Asian nations with the most stake in the Aral Sea. Despite this high-level exchange of views and the statements issued about the importance of solving the Aral Sea's problems, effective action has not yet followed.

Although the Karakum Canal permitted agriculture to spread across southern Turkmenistan and addressed the thirst of its cities, the unlined canal has also leaked water along its route for a half-century, dissolving salts in the ground that are then drawn to the surface by evaporation. Ironically, the irrigation canal increased salinity on 2 million acres to levels unsuitable for crops and fouled local wells near the canal.

The government hopes salinity on southern farmlands will be reduced by sending 2.6 trillion gallons of salty irrigation runoff water each year to a natural depression in northern Turkmenistan, creating Altyn Asyr (Golden Age) Lake. Begun in 2000, a 1,650-mile network of canals to be complete by 2020 will fill the basin until the lake reaches a surface area of 770 square miles and is 230 feet deep. The quality of the water in the resulting lake (perhaps better thought of as a sump) is a concern. But proponents claim that Turkmenistan's new man-made lake will serve as a habitat for migratory birds and a new source of agricultural water that will make the nearby desert lands bloom with plants and livestock.

The predicted benefits may be short-lived. Life tends to follow water, certainly, but how much transformation can happen in the desert will be limited by the salinity and toxicity of the runoff. The Salton Sea in Southern California also serves as a low-elevation collection point for runoff from Imperial Valley farms, but accumulated salts (plus fertilizers and pesticides) are now approaching concentrations lethal to fish, with effects passed along to millions of migratory birds that depend on the lake.

The billions spent (4.5 billion in U.S. currency) on the Golden Age Lake project could instead be used to finally line the Karakum Canal and return the conserved water to the Aral Sea, but this lake project has been a favorite scheme of Turkmenistan's previous two presidents (both dictators for life), who have embraced many grandiose Soviet-style construction projects.

Our goal that first day in the country was the city of Mary, which sits next to (and in some places sprawls into) ancient Merv, a State Historic and Cultural Park. Our guide was Yevgenia Golubeva, of Russian heritage, assigned by the national tourist bureau. One of the most important cities along the Silk Roads of Central Asia, Merv was an administrative,

Figure 10.   A few mud-walled buildings and this fortress still stand in ancient Merv, a Silk Road city in Turkmenistan.

trading, military, and religious center. Different cultures took turns dominating the city, each building its own mud-brick walls, houses, fortresses, and mosques. Greeks were in charge in the first few centuries B.C., backed by Alexander the Great's army. Under the Seljuk Turks from 1145 to 1153 Merv may have been the largest city in the world, with a population of 200,000 (Rosenburg, 2011). The Persian mathematician Omar Khayyam used an observatory there; he is perhaps most famous in our country for his poems published as *The Rubaiyat*.

Merv existed at that location because of water brought from a reservoir on the Murghab River. The Murghab flows from the mountains of Afghanistan and creates a lush valley, a welcoming contrast for settlers and travelers to the harsh Karakum Desert. The water that made the city possible was also its Achilles's heel, taken advantage of by sons of Genghis Khan. They arrived with an army of eight thousand soldiers in 1221 A.D. and were repulsed by defenders at the walls until the Mongols destroyed the city's dam on the river. With no water supply, the Mervians opened the city's gates to the besiegers, who proceeded to massacre the inhabitants. Merv's ancient glory came to an end.

Our tour, on a hot, sunny day, included the forts and mosques still standing; much of the ancient city's mud structures have dissolved away or been destroyed. We found cool shade inside a twelfth-century ice house, though its roof was missing; double walls still trapped an insulating dead-air space like a thermos, as nesting swallows flew in and out. Camels grazed in the historic park, though Turkmenistan had single-humped dromedaries, instead of the Bactrian two-humpers of western China.

On our return to Ashgabat, we visited downtown markets and city museums. We toured a stable for some of the nation's fabulous Akhal Teke horses, a type that may have been the source for the Arabian breed. The national museum's exhibit about the importance of horses in the Turkmen culture called them "the wings of a young man" and quoted a local proverb: "Getting up early in the morning, first greet your father, then greet your horse." A new horse-racing arena was adorned with statues and gleamed with marble and gilt trimming. Elegant stables housed the Turkmenistan president's retired horses.

Massive architecture is the rule for government buildings in Ashgabat, contrasting with Soviet-era cinder-block apartments on nearby streets. The image of the president appears in all public places. Ashgabat felt like some combination of *The Wizard of Oz* and a visit behind the Iron Curtain during the Cold War.

No other tourists were on the streets. Berkeli would like people from other countries to know that "we are kind, hospitable people, and would like to see more tourists. Though our economy is mostly based on minerals and cotton, there are lots of possibilities for tourism, which would provide people with jobs and an income. For example, Turkey's GDP is based 20 percent on tourism, but Turkmenistan's is only a fraction of 1 percent."

Authorities controlling travel into the country are not so welcoming. Paperwork for tourists is daunting, including an "invitation" required to enter the country before one can secure a visa and a sheaf of travel papers that one must carry everywhere. The bureaucracy seems to distrust foreign travelers. Before we left the Ashgabat airport, our bags were scanned at four different locations and our passports checked six times before we reached the waiting room.

The 38th parallel extends from Turkmenistan's western border across the south end of the Caspian Sea, which is about one-third as salty as the ocean. We flew above that great landlocked lake, the largest in the world, famed as a source of caviar. Seven species and subspecies of sturgeon have provided eggs for the delicacy, but the fish have declined to the edge of extinction due to overfishing and water pollution (pesticides, heavy metals, and oil), and because dams block spawning access up the Volga and Kura Rivers. Beyond the Caspian Sea, our plane passed over a small arm of northern Iran to reach eastern Turkey.

Hasankeyf in Peril on the Tigris River

Lake Van, in eastern Turkey, was our first stop on a 15-day trip across the width of that vast country, where the national government has built dozens of dams along its major rivers and campaigns have only just begun to modernize wasteful irrigation practices on farms that consume three-quarters of the nation's developed water.

A great inland sea in an enclosed basin, with alkaline water whose strange chemistry produces large calcium carbonate structures underwater, in a dramatic setting with snow-covered volcanic peaks and a lake surface thousands of feet above sea level . . . all of that sounds exactly like Mono Lake, our home on the 38th parallel. It also describes Lake Van, the largest lake in Turkey.

Lake Van also differs from Mono Lake in significant ways. It is much larger, with a 270-mile shoreline (Mono's is about 40 miles) and an average depth of 560 feet (today, Mono Lake averages about 60 feet deep), with its deepest point at 1,480 feet (the deepest place in Mono Lake is only about 150 feet).

Van holds a lot of water, but not so much salt. It is the largest soda lake on Earth, its chemical mix dominated by carbonates and also containing sulfates (both key ingredients in Mono), but without our home lake's chlorides—the ion associated with dissolved table salt. There are fish in Lake Van, though only a single endemic species is adapted to the unusual water chemistry—the Pearl Mullet (*Chalcalburnus tarichi*). Mono Lake is far too salty for fish. The alkalinity of both lakes is fairly close, around pH 9.8 for Van and 10 for Mono. Besides the mullets, Van supports hundreds of species of plankton and many migratory and nesting birds. Even its calcium carbonate structures are chemically related to Mono Lake's tufa formations. Van's are out of sight in the depths, where cyanobacteria build aragonite towers up to 130 feet tall.

The region's harsh continental climate (far inland from any ocean influence) brings cold, snowy winters. Lake Van is fed by rain and snowmelt and by the Bendimahi and Zilan Rivers, entering from the north, and the Karasu and Micinger Rivers from the east.

We arrived in the city of Van on the southeast shore of the lake (38°29′N). Adem, a student about to graduate from medical school, was on our bus. As many other helpful Turkish people would do during our sojourn, he introduced himself, answered all our questions, was interested in our project, and even walked with us through the city to be sure we found our hotel.

Following his directions, we took a minibus out to the lakeshore where local kids were swimming and women and their babies enjoyed the sun on the dock. The water felt slippery, and our wet fingertips tasted like baking soda. The water here, as at Mono Lake, can be used to wash clothes. Salts extracted from the lake are sold as detergents. Climate scientists have been drilling down through Lake Van's bottom mud to examine a half-million years of climate data from the sedimentary record.

Van was the capital of the Urartian kingdom between the tenth and eighth centuries B.C. In the first decades of the twentieth century, it was

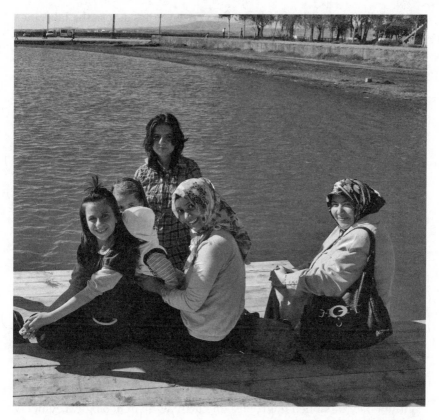

Figure 11. Women and children on a pier at Lake Van, a vast, alkaline sea in far-eastern Turkey.

at the center of the "ethnic cleansing" against Armenians that occurred when Turkey exerted control. Because the lake basin, surrounded by snowy peaks, is strikingly beautiful, an old Armenian proverb has it that "Van in this world, paradise in the next."

Today, desperate Kurdish refugees fill camps near Van to escape from political turmoil in Iran and Iraq. Far-eastern Turkey struggles with poverty and a history of ethnic strife between its Kurdish residents and the national government. Van was the only city on our journey where we felt threatened. That evening, a young man on a crowded street yanked Janet's sunglasses from around her neck and ran off.

Van is connected to the city of Tarvan, 60 miles across the lake, by a train ferry because a land route would have to contend with the lakeshore's extremely rugged terrain. We traveled by bus on the winding road near that southern shoreline, crossed the scenic Taurus Mountains, and descended through the upper Tigris River watershed to the city of Batman, a gritty oil production center. Our objective was a few miles farther south, the ancient village of Hasankeyf on the Tigris.

It was sunny and warm the day of our visit to Hasankeyf, a town with dwellings, chapels, and a fortress carved into limestone bluffs along the Tigris River (37°42'N). Our guide was Ipek Tasli, the young and energetic local coordinator for the Initiative to Keep Hasankeyf Alive, part of a coalition of seventy-two groups (municipalities, environmental organizations, and chambers of commerce). Hasankeyf is threatened by water that will back up behind Ilisu Dam, which is under construction a few miles inside Turkey's southern border. If completed, the 400-foot-high dam will create a massive reservoir and cover two hundred villages and displace up to 80,000 people. All but the highest sections of Hasankeyf would be submerged.

The setting, among rolling hills and sheer, honey-colored cliffs rising from the edge of the river, is breathtaking. This may be one of the oldest continuously inhabited town sites on Earth, with archeological evidence dating back 12,000 years. Hasankeyf served as a commercial center along the Silk Road from China during the early Middle Ages. Two stone piers and an arch are all that remain of the supports for a wooden bridge, built in 1116 A.D., that Marco Polo probably crossed.

We used a modern bridge, then hiked uphill to see some of the thousands of caves cut into the limestone cliffs for residences. Ruins of a hilltop castle built by the Byzantine emperor Constantine perched 600 feet above the river. Mongols destroyed most of the castle. A hidden stairway cut inside the cliff afforded access to river water during sieges. Hasankeyf eventually became part of Alexander the Great's kingdom and later was incorporated into the Roman Empire.

In a tea shop in one of the caves overlooking the main street, Ipek told us her work for the initiative had begun two years earlier, though the

Figure 12. Hasankeyf, on the Tigris River, one of the oldest town sites on Earth. Only the top of the minaret tower will remain above water if a proposed reservoir fills behind Ilisu Dam, downstream.

group formed in 1997. The Ilisu Dam is part of the Turkish government's Southeastern Anatolia Project, commonly referred to as GAP, from the Turkish-language project name: Güneydogu Anadolu Projesi. GAP is one of the largest dam and hydroelectric projects in the world. Planning on the $1.5 billion project began in the 1950s, and construction in the 1980s. Twenty-two dams are being built along the Tigris and Euphrates Rivers, along with nineteen hydroelectric power plants. The water is projected to double Turkey's irrigated farm acreage.

Our recently published maps were already out of date in depicting new reservoirs. Few free-flowing stretches remain on the rivers. As we traveled by bus farther west, passing one reservoir after another, each dam backed up slack water almost to the base of the next.

International resistance to the Ilisu Dam project developed because the reservoir will flood extremely significant historical and archaeological sites. Swiss, German, and Austrian underwriters within the European

Union that had funded initial work pulled out in December 2008 after so many objections were raised. A list of 150 World Bank conditions dealing with the environment, protection of heritage sites, relocation of residents, and effects on Iraq would have to be addressed if the European underwriters were to provide funding. The Turkish government, however, vows to fund the construction on its own and projects completion in 2017. Ipek's coalition wonders if hidden money, possibly from China, is making that possible. The local Kurdish population is wary, considering this dam puts additional pressure on the region's Kurdish communities to assimilate into the Turkish culture. As earlier dams were completed as part of the GAP project, many people forced to resettle received no compensation from the government because land titles were too poorly recorded to back up claims. Kurdish villagers moved away from rural areas to urban centers, especially the cities of Batman and Diyarbakir, where they had to develop new job and social skills to survive.

"Here is not Turkey," Ipek told us. "Of course, it is in the territory of Turkey, but we are not Turks. We are a different nation [that includes] northern Iraq, part of Iran, and part of Syria." She worries about being arrested for opposing a state project, which is a crime in Turkey.

Turkey's dam building and the reduced flows downriver on the Tigris and Euphrates are generating tension with neighboring Syria and Iraq. The two downstream nations demand that more water be released, but Turkey remains focused on filling its reservoirs. Forty percent less water will flow to Syria and even less to Iraq, farther downstream. Both nations depend on the Tigris and Euphrates to grow food and serve growing populations. Here is yet another stressor to the political stability of the Middle East.

Despite these concerns for the future, the mood in Hasankeyf was celebratory on the day we visited, which was a national holiday. Returning to the riverside, we watched people of all ages splashing in the shallows and young adults dancing and singing in the street. Seated on divans around a low table in a restaurant built over the flowing water, we ate a meal of fish, taken fresh from the river and barbecued, while looking down at other fish nibbling at our crumbs.

The Tigris River creates a unique riparian ecosystem. The local economy is based on fishing and farming. If the reservoir is filled, conditions for native fish will change and the river ecosystem will be obliterated. A Turkish environmental group, Doga Dernegi, has made a local soft-shelled turtle (*Rafetus euphraticus*) the symbol of these consequences. The turtle used to be found along the Euphrates River, but is nearly gone there now because so many dams have been erected. Ilisu Dam may bring the species to the brink of extinction. Turkey has no law equivalent to the United States' Endangered Species Act, nor, apparently, a public trust doctrine on which to rely for legal protection. It could help if Turkey was accepted into the European Union, which has more progressive environmental laws.

That some people seem to care more deeply about these environmental concerns than about the likely impact on the local people amazed Ipek. Twenty-five years old, knowledgeable and eloquent, she told us that a German ambassador to Turkey has said that the nation is big, has lots of other cultural sites, and Hasankeyf's earthen cliffs and caves will be destroyed by time anyway. Despite such unfortunate "If you've seen one ancient site, you've seen them all" attitudes, she and her group remain determined to lobby internationally for pressure on the Turkish government. "Hasankeyf does not belong to Turkey, it belongs to humankind," Ipek said, her dark eyes blazing. "We ask the people of the world to save Hasankeyf."

We stood knee-deep in the Euphrates River after a long bus ride through the hills of eastern Turkey, westward from Hasankeyf toward the city of Malatya. The Euphrates is the longest river in southwestern Asia, rising in highlands northwest of Van and terminating far to the south in the Persian Gulf. The fertile land between the lower Tigris and Euphrates Rivers cradled Mesopotamian cultures, and water from both rivers enabled the first irrigated agriculture, nurtured some of the world's oldest known human settlements, and inspired the dreams and follies of kings. Only a few miles to the east, archaeologists were uncovering a 7,000-year-old city called Arslantepe.

North of Malatya, we waded into the warm water at a city park beside a ferry dock. On the way there, the local bus passed through Battalgazi, the site of an ancient caravanserai for Silk Road travelers and their camels. The Euphrates was backed up there in yet another GAP reservoir, the Karakaya. Tamarisk trees lined the shore and provided shade in the park. They are an exotic problem species for desert wetlands in the United States, but part of the natural order in Central Asia.

Suddenly we realized that we had attracted a half-circle of *gendarma*, the local military police, all carrying machine rifles. "What are you doing?" one asked. The English speaker was a handsome young man doing three months of national military service (we would learn) who taught English in Malatya and had been an exchange student in Seattle, Washington.

"Is it a problem to be here?" we asked. The answer was no, and though their demeanor seemed friendly, being confronted at the edge of the water by rifle-toting soldiers made us uneasy. We exchanged a worried glance, wondering if going to places not frequented by tourists was a mistake that had finally caught up with us. Pulling out our map, we explained the book project that had us traveling around the world along that local latitude line.

The soldiers were there to meet a passenger ferry, which arrived at the dock a few minutes later. One held the leash of a narcotics-detection dog that splashed into the water and lapped a drink. The ferry provided local service; internal security is a priority in this Kurdish section of Turkey where tensions have led to violence.

Once the soldiers finished inspecting the arriving passengers and vehicles, the one who had spoken to us returned, saying that his commander invited us to come to their barracks in Battalgazi, have tea, and talk about their area's historic sites. We were not sure what to think of the invitation. Was declining really an option? We explained that we were using the public bus system.

They told us they could arrange a car ride for us. We walked with the translator to a sedan in which two men in civilian clothes sat in the front seat. They had been in the park when we arrived, perhaps undercover. Neither of those men spoke a word or even turned to look at us.

Our car raced by the slower truck full of soldiers. Again, we looked at each other, silently wondering just what was coming next. The car passed through town and then through a gate into the Battalgazi Gendarma Komutanligi compound. After the gate swung shut, we were politely escorted to a nearby patio and served tea and mulberries. The commander, another officer, and our translator-soldier joined us. (We asked everyone's name and also whether we could take their photographs, but were allowed only to record the name of Commander Samim Paksoy and that of the barracks.) They were gracious people, quite interested in our book project. On a map, we showed them our travel route across their country, and they pointed to their hometowns in central Turkey. They were proud of the nearby Arslantepe archaeological site and presented us with a book about mound excavations there, where an Italian group has been at work for 40 years.

Our worries proved a bit ridiculous, yet it was a big relief when our translator escorted us to a nearby bus station. When the public bus pulled up, he spoke to the driver about dropping us near our hotel in Malatya, and when we arrived, the driver would not accept any money for the fare.

．　　．　　．　　．　　．

Malatya is famous for its apricots and the colorful bazaar where they are sold. Kamal, an ageless, effervescent character with long curly hair around a bald dome, holds court in his travel "office"—a table at an outdoor café in the park—and arranges tours to Nemrut Dagi.

On that 7,000-foot mountaintop south of Malatya, a king named Antiochus I (who ruled Commagene, a kingdom bounded by the Euphrates River) built a monument to Roman and Persian gods in the first century B.C. and placed his own statue among them. Fine sculptors shaped the likenesses of Apollo, Zeus, Hercules, Tyche, and of course, King Antiochus himself. The magnificent sculpted heads were forgotten by history until they were rediscovered in 1882 by an archaeologist. Major restoration was completed in the early 1980s.

Nemrut Dagi is one of Turkey's must-see treasures, though it takes considerable effort to reach the mountain. We opted for an overnight trip to a lodge nestled a mile below the summit. The remote summit of Nemrut Dagi National Park lies almost exactly on our line, at 37°59′50″N.

We visited the mountaintop terraces and sculptures at sunset and again at sunrise. The massive figures once loomed 25 to 30 feet tall. Roughly hewn bodies still stood, but heads had fallen from above before they were unearthed. They still have a captivating, mystical quality. Honey-colored stone reflects the changing light, and the setting, on a conical mound crowned with small stones that may cover burial chambers, is magnificent. Shining through the morning mist, another reservoir could be seen far to the south, formed by Ataturk Dam on the Euphrates River—a reminder of the water source that irrigated the crops that generated the taxes that supported Antiochus's kingdom and made his extravagant monument possible.

Fairy Chimneys, Tuz Golu, Travertine, and a City That Lost Its Port

In the Cappadocia region of central Turkey, a full day's ride from Malatya on local buses, volcanic ash hardened into rock and eroded by water through the ages has created a fantasy landscape. We left the high mountains and river canyons behind, dropping down to agricultural valleys before following a twisting road into the fairy chimneys, towers, and mushroom shapes that surround the village of Goreme. In the distance a huge snow-capped volcano stood sentinel.

The whitish rock formations are locally called tufa. In our many years of telling visitors about Mono Lake's tufa towers—the photogenic limestone formed when springwater and lakewater mix—we had often heard about "tufa" formations in Cappadocia, though, technically, they are volcanic tuff. Layers of ash that consolidated into tuff rock were gradually eroded by flowing water and windblown sand. The ridges, columns, and freestanding towers that resulted were capped by a more resistant, darker layer of basalt, yielding mushroom shapes and distinctive "fairy chimneys." Residents carved cave houses and chapels into the formations, and many are still occupied, with gardens and apricot orchards adding to the Hobbit feel of the valleys. Visitors may even choose to stay in cave hotels complete with windows, terraces, and satellite dishes. Entire underground cities in multiple layers provided refuge during long-ago raids, and even horse stables had been carved into some of the larger rocks.

Three days was barely a beginning for exploring the lovely area. Most of our time was spent hiking, seeking the elevated viewpoints for looking down and across the valleys. Many tourists seek similar views in the dawn balloon rides that contribute significantly to the local economy. Wildflowers abounded, with Turkey mullein (another exotic pest at home that is native here) and irises decorating the hillside. A fox showed itself for a moment. Camels were there too, as a reminder of Turkey's historic connections with China via the Silk Road.

From Goreme, we made a day trip to find Tuz Golu, its name meaning "salt lake." Konya is the nearest big city in that hydrologically closed basin, a central Anatolian farming region renowned as Turkey's breadbasket (37°52'N). Some of the world's thirstiest crops are the specialties here: wheat, sugar beets, cotton, and corn. Turkey is the seventh-largest cotton and sixth-largest sugar beet producer in the world, yet less than one-tenth of it farmland uses modern irrigation techniques. In 2009, pilot projects developed on several hundred acres in Anatolia by World Wildlife Fund–Turkey installed drip irrigation and trained farmers to use the new system; the projects achieved 37 to 50 percent water savings.

Driving north toward Tuz Golu from Konya, we paralleled canals carrying the primary twenty-first-century source of water to the massive inland sea: treated sewage from the city. A huge white expanse shimmered in the heat, but it proved surprisingly hard to reach the shore of Turkey's second-largest lake. From its southern end at 38°30'N it extends north for 50 miles. "Hypersaline" is barely descriptive enough; at 33 percent salinity, this is one of the saltiest lakes in the world.

The water that filled Tuz Golu that spring was fast evaporating with the warm temperatures. The lake appeared tantalizingly close, but reaching the "shore" required a long slog through farm fields and wildflowers, with the water itself still far across Tuz Golu muck. In winter, the lake's deepest point is only six feet down. Every summer it goes almost completely dry, leaving behind a thick layer of salt that is commercially harvested and provides 64 percent of the nation's salt supply. Pollution entering the lake had been making its way onto dining tables via that salt, so a wastewater treatment plant, completed in 2009, had become an urgent health matter.

Despite its expanse, the lake's high salinity limits life. There were no waterbirds to be found that day, though the largest Turkish breeding colony of Greater Flamingos (*Phoenicopterus roseus*) nests on a group of islands in the southern part of the lake. Their numbers were low the year of our visit, but rebounded in 2011 to a record 16,000 young in a breeding season that followed heavy rainfall in central Anatolia. Plans to increase salt harvesting continue to be a threat, along with global warming and diversions of streams and groundwater to farms in the watershed.

We could see the lake's islands, doubled by their reflections on the water. Land birds did appear as we walked: storks and harriers flew overhead, while many small birds flitted in the fields. A startled jackrabbit bounded out of the grass. We wondered when it had last seen a human visitor. Our approach to the lake was definitely not developed for visitors, though Tuz Golu's water and surrounding wetlands have been declared a specially protected area.

Leaving Cappadocia's strange landscape, we boarded another public bus and headed west on modern highways that follow the ancient Silk Road caravan route into western Turkey. Some caravanserai locations—those "truck stops" for camel caravans—still serve transport and travelers. The driver stopped for gas in Sultannan, 3,000 feet above sea level, where a caravanserai was established in 1229 A.D. Above the high plain we were traveling, Mount Hasan towered in the distance, a conical snow-capped volcano 10,673 feet high, and one of the regional sources of the volcanic deposits that make Cappadocia so magical.

Turkey has the largest fleet of bus companies in the world, needed in a large country where few people own cars and gasoline is expensive.

In the east, buses were older than in the western tourist zones, ran less frequently, and were more crowded. Even the older buses, though, were surprisingly comfortable. There was always a steward on board offering water, tea, and lemon hand freshener. The air conditioning usually worked. In the west, buses were fast and immaculate with nicer snacks.

Bus depots were clean, complete with young men or boys that greeted you with a cup of tea as you stepped off. (Tea is ubiquitous in Turkey. The famed Turkish coffee is rare and expensive.) At bus stops, men smoked outside, while the women disappeared into the restroom, where (Janet learned) headscarves came off and everyone lit a cigarette, away from the gaze of men.

One memorable ride began with a huge refrigerator being loaded under the bus. Later, the driver stopped at a roadside market where twenty cans of gasoline were laboriously poured one by one into the gas tank, while passengers smoked cigarettes a few feet away.

Tufa towers are the signature visual element at Mono Lake, so it was interesting to learn that the 38th parallel in western Turkey coincides with spectacular features made of calcium carbonate—the travertine terraces of Pamukkale National Park (37°55′N). Brilliant white terraces and aquamarine water draw tourists from around the world. The face of the travertine hillside looks out over orchards and wheat fields near the Büyük Menderes River (which twists so much it has given us the word *meander*). Lukewarm springwater at Pamukkale is loaded with calcium carbonate, which precipitates as travertine when exposed to air, in a process similar to cave stalactite formation. It is another example of how calcium, carbonate, and water combine to form beautiful features around the world.

The white terraces must be dry to really gleam, so park staff shift the flow of warm springwater (95° Fahrenheit) throughout the day, which also provides changeable bathing experiences for visitors. The low-tech system includes ditches, sheets of metal, and plastic bags. The challenge for Pamukkale's managers is to protect the terraces while accommodating thousands of bikini-clad tourists. People are allowed to swim and frolic in artificial pools constructed along an old road routed through the

Figure 13. Pamukkale's travertine terraces form where calcium carbonate precipitates from warm water in western Turkey.

natural terraces. A crew armed with brooms was cleaning dirt and algae from the area that was off-limits.

Adding to the scene are spectacular Roman ruins above the travertine hillside, where Heirapolis was founded in the second century B.C. as a thermal spa and healing center. Multiple earthquakes eventually forced the Romans to abandon the city. A well-preserved ancient theater overlooks Roman baths where modern visitors still float above broken marble columns on the bottom of the pool.

Our day in Pamukkale was an interesting mix of wading, people watching, and archaeology, punctuated by the shrill whistles of guards whenever people ventured too far up the terraces or neglected to take their shoes off—bare feet are required for everyone, including the uniformed "rangers." The white hillside was lit up after dark, as frogs in the lake below croaked and warbled songs unlike any we had ever heard from the frogs we knew.

At the west coast, Turkey's grand archaeological jewel is Ephesus, which nestles in a canyon in a pine forest at 37°56'N. In its heyday, between 100 and 200 A.D., 250,000 people lived in this Roman capital of Asia Minor, making it the world's fourth-largest city after Rome, Alexandria, and Antioch. Notables such as Cleopatra, Alexander the Great, and Saint Paul are tied to the history of this great city. The founders of this Greek harbor town, created a man-made port at the mouth of the Cayster River, which kept silting in, requiring the Greeks and then the Romans to continually dredge the navigation channel. As part of the Byzantine Empire between the fourth and thirteenth centuries, the Ephesians stopped maintaining that critical tie with the outside world, and river sediments pushed the shoreline several miles away. Ephesus had to give way to Samos as the key port in the region.

Today, from the top of the huge amphitheater the sea is barely a gleam in the distance, five miles away. Hints of the city's watery history remain, however, with a Harbor Boulevard (paved with marble), harbor gymnasium and baths, and wetland marshes where the port used to be. At the edge of the marsh, on an old church site, several little frogs hopped across the stone floor.

The Celsus Library of Ephesus, built in 117 A.D., is breathtaking, even when crowded with tourists from the cruise ships that dock in Kusadasi. Terrace houses replete with frescoes and mosaics are being pieced back together like giant jigsaw puzzles. Our new Turkish friend, proprietor of the Liman Hotel beside the modern port, said that what impressed him most about Ephesus was how its ancient residents constructed buildings "without any sticky stuff" (mortar) or modern measuring instruments.

Figure 14. Ephesus, on Turkey's west coast, was a Roman seaport until the river mouth silted in. Today the coast is miles away, barely visible from the top row of seats in the ancient theater.

One of the Seven Wonders of the Ancient World is the nearby Temple of Artemis, just outside Selçuk, which served travelers stopping to pay homage to the goddess on their way to Ephesus. An early version of the temple, constructed about 550 B.C., was destroyed by fire, and a grander replacement begun in 356 B.C. It was still under construction in 333 when Alexander the Great came to Ephesus.

About 140 B.C., Antipater of Sidon named the world's seven wondrous sites, in a list that praised the Temple of Artemis above all the others:

> I have seen the walls of rock-like Babylon that chariots can run upon, and the Zeus on the Alpheus, and the Hanging Gardens and the great statue of the Sun, and the huge labor of the steep pyramids, and the mighty Tomb of Mausolus; but when I looked at the house of Artemis, soaring to the clouds, those others were dimmed. Apart from Olympus, the sun never looked upon its like.
>
> Antipater, XCI, 69 (Gow and Page, 1968)

Only one 60-foot-high column remains to hint at the 127 marble columns in that great temple (the Parthenon in Athens has 86 columns). The single column remaining incorporates dissociated fragments discovered on the site in the twentieth century. Storks were nesting on top when we were there.

Artemis was a fertility goddess for the people of Ephesus. Her statue in the museum in Selçuk has multiple breasts (or they may have represented eggs or even bull testicles, according to the museum staff). The site is marshy; Pliny the Elder wrote that such wet ground was preferable as protection against earthquakes. In 391 A.D. the Roman emperor Theodosius the Great closed the temple after declaring Christianity the new state religion.

The contrast between hordes of cruise ship tour groups at Ephesus and the lonely Temple of Artemis was striking. We wandered alone in the marsh at the end of an unmarked road that once held one of the wonders of the ancient world. Imposing monuments can, it turns out, have fleeting lives, making those that still remain along the marbled streets of Ephesus that much more precious.

Our westward progression through Cappadocia and Pamukkale had shown us the role of water in spectacular natural landscape features and historical sites that are major attractions for travelers. Until Cappadocia, we had seen almost no other international tourists since leaving the Terracotta Warriors in Xi'an, China, thousands of miles to the east. We had crossed 750 miles of inland Turkey by this time, and a dominant impression was how surprisingly green the country appeared. In our ignorance we had envisioned Middle Eastern deserts. Instead, almost all of the land out the bus windows was irrigated and cultivated, except where highways ascended mountain ridges. Turkey is bountiful enough to feed its population and still have extra produce and grain for export, though the country's key rivers have been tamed at the expense of history and nature and a tremendous waste of farm water could be saved by reforming irrigation practices.

Across Asia, from Korea to Turkey, many water problems we encountered remain unresolved. The South Korean project to "restore" its four major rivers was racing toward completion, at the expense of river eco-systems, riparian vegetation, and farmland. No effective measures have been taken to reduce river diversions in Turkmenistan to help the Aral Sea. Although international voices speak out against Turkey's plan to flood ancient Hasankeyf, that government still continues construc-tion on the Ilisu Dam on the Tigris River. Tensions between Turkey and nations downstream that share the Tigris and Euphrates heighten concerns about peace in the Middle East. Yet we were impressed by the energy and dedication of activists and park professionals in those nations, even in Turkey, where activists risk arrest when they oppose official government programs.

Our route now entered the Mediterranean, a region that shares cli-mate patterns and a number of water challenges with our home state of California.

PART TWO Europe

Never think that a fight cannot be won. Never, never.

Anna Giordano

Map 2.   Europe
1. Izmir; 2. Delphi, Greece; 3. Messina; 4. Mt. Etna; 5. Sierra Morena;
6. Guadalquivir River

Greek Islands, Athens, and the "Navel of the Earth"

The west coast of Turkey borders the Aegean Sea and Greece, a nation whose boundaries encompass thousands of islands—more water than land. Even on the mainland, one is never farther than one hundred miles from the sea. Ferries are the key transportation mode connecting the Greek islands. Most serve cars and trucks along with passengers and feel like small cruise ships, capable of smoothing out the Mediterranean's wind-tossed waters. High-speed catamarans, solely for passengers, are also available, with indoor seating that resembles the cabin of a very wide airplane.

As one approaches them on a ferry through the Mediterranean Basin's perennial haze, many islands appear to be capped by white snow that soon transforms into whitewashed buildings perched on the cliff tops. Although the Mediterranean Sea is often an intense turquoise blue under the 38th parallel sun, the sky we saw was usually white. We wondered if the haze was simply seasonal humidity, but learned that particulate pollution travels south from Europe to settle in the basin. Throughout the Mediterranean Sea region we were struck also by a lack of sea and bird life, compared to the California coast. The sea is seriously overfished, from shrimp species, anchovies, and sardines, to the top predators, Bluefin Tuna and swordfish. Much of the seafood common on restaurant menus must now be grown in aquaculture "farms," which specialize in Sea Bass and Sea Bream. Jellyfish "blooms" have become common and widespread, a nuisance to swimming tourists and yet another predatory pressure on fish. While snorkeling, we suddenly realized we were surrounded in every direction by large globular jellyfish, beautifully purple, but with menacing tentacles. Jellyfish may be taking advantage of the ecological vacuum in which other predators have declined, but warming Mediterranean seawater—another consequence of global climate change—appears to be enabling the spread of these creatures.

People have applied pressure to this environment for a very long time, and it shows.

A Turkish ferry from Kusadasi took us through the back door to Greece via the island of Samos (37°46'N). Along the town waterfront, a door stood open to the Samos Ecological Society. Inside, we met three very friendly, helpful people, though only Michaliadis Michael spoke English. The others kept smiling and nodding and answered our questions as Michaliadis translated. On the wall were posters about World Wetlands Day and also about Monk Seals (*Monachus monachus*).

The group works on education and environmental issues, particularly the rare, extremely endangered seal. Samos is one of the Monk Seal's last strongholds. Declining fisheries affect the seals' food too, and they are sometimes directly killed by the illegal use of dynamite by fishermen (one dead seal had washed up a few weeks earlier). There are also concerns about uranium in the seals' habitat, left by military missile exercises.

Exploring the west end of the small island, we hiked above a seal protection zone, scanning with binoculars for little heads in the sea. Alas, the Monk Seal remained elusive.

On the south side of Samos Island, near the village of Pythagorio, the Eupalines water tunnel was cut through a mountain more than 2,500 years ago. The Mediterranean climate pattern, with winter rain and a long summer drought, creates water supply issues across Greece (and the entire Mediterranean region) that are familiar to Californians. This 3,432-foot water tunnel was hand-chiseled through the mountain to bring water to a thirsty city of 80,000 people (today, the total island population is 32,000, with the biggest towns on the north coast). Working from both ends, the two crews met unerringly in the middle. In 524 B.C. they had no surveying equipment or compasses, but rather used a series of right triangles to calculate distances and stay on course, though Pythagoras (whose theorem about right triangles is familiar from high school geometry) was not born here until many years later. These were major mathematics and engineering feats.

What an intriguing connection for us! A thirsty city reaches out and diverts surface water into conduits that pass through a tunnel bored beneath a mountain, then through an aqueduct to reach the city—a phenomenal achievement for its day and an example of just how much toil some leaders could set in motion (on Samos, by a despotic ruler, Polycrates, who used slave labor) to solve a perceived water supply problem. History repeated itself near our 38th parallel home when the city of Los Angeles reached into the Mono Lake Basin to divert streams and avoid the limits of the city's own natural watershed. It passed that water along to the Los Angeles Aqueduct through an audacious engineering feat, tunneling 11 miles beneath the Mono Craters mountain range.

In both cases, the efforts were a response to water challenges that go with a Mediterranean climate, characterized by wet winters, but long summer droughts. Found on only 2 percent of the Earth's land, between latitudes 30° and 45° in both hemispheres, Mediterranean climates are confined to the southwestern edges of continental land masses, including South America, Australia, South Africa, and California, besides the sea basin that gives the climate type its name.

The nearest Greek island to Samos is Chios, to the north (38°12'N). Ample groundwater reserves on Chios include springs that gush year-round but are no longer drinkable. Overdrafting the groundwater to serve summer-season demand at this popular tourist destination allowed seawater to intrude into the aquifer, so now most of the island's water is suitable only for washing. On most of the small Greek islands, wells are the key source of freshwater, as there are few surface streams or lakes to tap. Residents capture rainwater in household cisterns, but not nearly enough to serve the seasonal invasion of summer tourists. Many Greek islands have resorted to paying for freshwater delivered by tankers at enormous expense.

That expense makes the cost of desalination a feasible alternative. Plans to desalinate salt-contaminated springwater are under way on Chios, while other islands in the Aegean Sea are considering the greater challenge of seawater desalination. Costs drop, as do greenhouse gas emissions, when the energy needed to force saltwater through reverse-osmosis membranes can be generated by wind turbines. Floating offshore facilities are preferred because space is limited on small islands and wind speeds are generally higher on the open sea. If facilities can be positioned near currents to disperse the salty brine left after freshwater is extracted, the impact on the ocean ecosystem may be acceptable, though that remains a concern. The first floating, self-powered desalination units have begun testing off the island of Iraklia, farther south in the Aegean Sea.

The ancient city of Athens is one of the largest cities in the world along the 38th parallel (the famous Acropolis in the city center is at 37°58′N). From Athens, our mainland travel route would circle the Gulf of Corinth, taking in ancient Delphi, the "navel of the Earth"; the Peloponnese site of Olympia, home to the ancient Olympic Games; and the city of Corinth.

Athens was founded around the Acropolis, on which steps to the Parthenon have been smoothed by thousands of feet for thousands of years. The Acropolis, like many Greek hills, is porous carbonate rock through which water can dissolve its way along fissures and form caves, sinkholes, and natural "pipelines," often emerging as springs farther downhill. Springs and seeps from the limestone Acropolis were used by Mycenaeans as early as the thirteenth century B.C. Anthropologist Brian Fagan has likened the Acropolis to "a huge reservoir" (2011, 162).

The annual summer droughts, a seasonal challenge faced across the region, also posed water problems for the ancient city as it grew. Cisterns stored water and fed street fountains, some of which are still used. Once local wells could no longer handle the demand, aqueducts were built to transfer water from nearby mountains. In 140 A.D., during Roman rule, the Hadrian Aqueduct was completed from the foot of Mount Parnitha to Hadrian's Reservoir. As in Samos, routing aqueducts underground was intended to protect their precious supplies from enemy attacks. Hadrian's Aqueduct included a 15-mile-long tunnel chiseled through rock, with air shafts at approximately 100-foot intervals—a tremendous feat for slaves using hand tools. Pipes distributed water through Athens, and drains handled sewage and storm runoff.

When Athens absorbed several million refugees from Turkey, in 1923 at the end of the Greco-Turkish War, the city's water needs became critical. About 26 miles to the north, along a route that today passes through planted fields, vineyards, olive groves, and lush hillsides, is the city of Marathon. On the Marathon plain in 490 B.C., the Greeks defeated an invading Persian army. That battle location is remembered today for the long-distance run made by Pheidippides, the messenger who delivered news of victory to Athens and then died of exhaustion. Near Marathon, a reservoir was completed in 1929 that provided most of Athens's water supply into the 1950s. The same construction project included renovations to the ancient Hadrian Aqueduct system, which still served the city. Marathon Dam, which backs up water from the Charadros and Varnavas Rivers, is uniquely surfaced with white marble from the same source used on the Parthenon.

A more recent, post–World War II population boom moved Athens to add Yliki Reservoir to its supply sources in 1959, using pumps to move water across hills northwest of Marathon. In 1981, the Athenians reached 120 miles farther west to the Mornos River and built there one of the tallest earth gravity dams in Europe (375 feet high).

The landscape and vegetation northwest of Athens, north of the Gulf of Corinth, looks much like California's coast range. Blue-domed churches in each village and signs in the Greek alphabet were the primary giveaways that this was Greece. The country is hilly with many canyon

Figure 15.   The Temple of Athena at Delphi was home to the Greeks' prophetic
oracle. The ancients considered Delphi the "navel of the Earth."

creeks that go dry in August. Away from the capital city, ancient agricultural lands support olive orchards, vineyards, and irrigated vegetable
crops. Olives and grapes are crops well adapted for the Mediterranean
climate, but hot, dry summer days present familiar water challenges.
In Greece, 85 percent of water consumption is by farms, a higher per-
centage than in most nations. Although here, as in Turkey, the World
Wildlife Fund has been working to encourage less wasteful irrigation
practices, the national government has focused on diverting water from
the Archeloos River, in western Greece, and building yet another long-
distance aqueduct to the central farming belt.

Through the ages, water has played a special, mystical role in Greek
culture. The ancient Greeks associated natural fountains and springs
with gods and goddesses and extolled their medical or mystical powers.
A perfect example is Delphi, south of Mount Parnassus, where we
learned that "the navel of the Earth" is a feature of the 38th parallel
(38°29'N).

At Delphi's Temple of Apollo, a sacred spring called Castalia flows into the Cephissus River. Castalia was one of the naiad nymphs, goddesses who, according to Greek mythology, inhabited and protected rivers, streams, lakes, marshes, fountains, and springs. The nymph Castalia aided the Delphic Oracle. Pilgrims posed questions for the oracle priestess, who sat beside a cleft where sweet-smelling vapors rose from a spring below ground. She entered a trance, then recovered to deliver answers from Apollo, the god of prophecy, healing, and the sun.

Modern analysis of the spring in the sanctuary has detected ethylene. In low concentrations the sweet-smelling gas can induce a trance state, but exposure sometimes causes violent reactions, including wild, incoherent cries and thrashing—behaviors that fit historic descriptions of the priestesses' sometimes erratic behavior. Christian emperor Theodosius I, who shut down the Temple of Artemis in Greece, also ended operations at the "pagan" sanctuary at Delphi in 395 A.D.

Northwest of Delphi are the Messalongi marshes, the largest wetlands in Greece. We vowed to return some winter at the peak of the bird migration. The largest natural lake in Greece, Lake Trichonis, is to the north. Given that California is struggling with problems posed by non-native zebra mussels introduced to lakes and streams, it was interesting to learn that this lake is part of the mussel's native habitat.

The bus trip between Delphi and Patras, the port city at the western mouth of the Gulf of Corinth, followed the shore so closely that our bus tires were in seawater when waves swept across the beach road. Patras is Greece's second-largest city and a major port, roles played by San Francisco in our home state. It even has a dramatic bridge spanning the "gate" into the gulf, called the Rio-Antirrio Bridge (38°08′N). Rows of cables on the recently completed bridge stretch like harp strings from four towers. We had our cameras poised for the crossing, but just before we reached the bridge, our bus turned down to an old ferry terminal. Puzzled, we asked why and learned that the toll is three times higher on the new bridge, so public buses still use the ferry.

Since we had left Athens, olive trees had covered nearly every hill, but south of Patras flat farmland grew corn and melons. Not far to the west, in the Ionian Sea, the island of Zakynthos (37°47′N) has six beaches that serve as the most critical nesting areas for endangered Loggerhead Sea

Turtles (*Caretta caretta*) in the Mediterranean. Volunteers with Archelon, the sea turtle protection society of Greece, work along those beaches during the nesting season, from early May through mid-August, educating some 70,000 tourists each year about the turtles' need for undisturbed beach sand. Loggerheads migrate long distances to reach the island and come ashore at night to lay eggs. Key challenges for Archelon lie in tourism development and making sure laws are enforced that limit hours of beach activity and the numbers of beach umbrellas and sunbeds. Since the turtles use the back edges of the beaches, the law limits human activity to the beach sand closest to the water. The organization even convinced the government to stop night flights from the island airport, because the noise and lights confused turtles on a nesting beach in the flight path. It successfully lobbied for the establishment in 1999 of the National Marine Park of Zakynthos, and now operates a sea turtle rescue center at Glifada, south of Athens, that treats injured turtles and serves as an environmental education center.

Turning away from the coast, our bus climbed back into hilly country on the Peloponnese Peninsula toward Olympia, site of the ancient— the original—Olympic Games (37°38′N). At that remote location, where the Kladeos and Alfeios Rivers meet, we finally saw a truly blue Greek sky. The ancient stadium sits quietly, enlivened only by tourists posing for photographs as if starting a race on the track; the site's tranquility is interrupted only when torch-lighting ceremonies return every four years to launch the modern Olympic Games. Greece was not united when the original Olympics were held (776 B.C. to 393 A.D.). When each athletic competition was held, a truce was declared between participating warring city-states. A 38-foot statue of Zeus made of gold and ivory, erected at Olympia, became yet another of the seven wonders of the ancient world along the 38th parallel (though it disappeared toward the end of the fourth century A.D.). Dedicating the Olympic Games to Zeus did not sit well with Christian emperor Theodosius I, of course, so he ordered the events ended.

Back on the bus and heading east toward Corinth, we felt our ears pop as we wound up and down rolling hills, following an ancient highway route through the Alfeios River canyon and a landscape reminiscent of the Sierra Nevada foothills. In Greek mythology, the Alfeios was one of two rivers Hercules rerouted to cleanse the Augean Stables. The road narrowed where it passed through old towns, and whenever oncoming traffic met our bus, horn blowing and complex maneuvering ensued to decide which vehicles would have to back up and pull aside. The length and winding curves of this highway help explain why this part of Greece remains undeveloped and little disturbed by tourism. In a place where an unoccupied car blocked the road, our driver's horn brought a woman from a nearby house. She came to his window, spoke awhile, then walked back into the house to find her keys, cheerily waving at the people on the bus. There was no grumbling while we waited; a lady so exceedingly pregnant is accorded great respect in Greece.

The Corinth Canal cuts through the narrow isthmus that separates the Peloponnese from the Greek mainland, providing a shipping shortcut from the Gulf of Corinth to the Aegean Sea. In 67 A.D., Roman emperor Nero put several thousand workers to work digging, but the canal was not completed before his death. At that point the work stopped and did not resume until 1881. Finished two years later, the 26-foot-deep, four-mile-long canal is only 79 feet wide, too narrow for many modern ships, yet still a busy conduit for commerce.

We were met at the Corinth bus station by Nico, who keeps an apartment for tourists beside his house and runs a tavern in town. He drove us up the hill to the Acrocorinth fortress, the oldest on the Peloponnese (37°54'N). This had become a notorious site by the time Paul's first letter to the Corinthians, now in the Bible, warned the citizens against visiting harlots there. Fortress walls show modifications by a succession of outside rulers: French, Byzantine, Venetian, and Turkish. At the summit, there was an expansive view of the Gulf of Corinth, which we had circled during the prior week. Greenhouses dominated the area between the hill and the sea; Corinth is especially famed for the currants grown there.

The Acrocorinth is, like the Acropolis, a porous limestone hill with many springs emerging along its northern base. Legend has it that a woman named Peirene was transformed into one of those springs by tears she shed after her son was killed by Artemis (bringing to mind that goddess's temple in Turkey). The ancient gods and goddesses spent considerable time, apparently, along our line of travel.

Back in Athens, on our last night in the country, we checked into a hotel after dark and walked to the Parthenon. Lit up, the ancient temple glowed high above this extraordinary city on the 38th parallel.

In Italy, the 38th parallel passes through the south end of the mainland, then closely follows the north coast of Sicily from Messina to Palermo. A ferry ride from Calabria brought us to the fabled island across the Strait of Messina.

Saving Migratory Raptors and Drinking Water in Sicily's Mafia-Dominated Culture

The narrow Strait of Messina funnels raptors seeking the shortest route across the Mediterranean from Africa during their annual migrations. Migrating birds travel south to north and then back again, across lines of latitude to seek "endless summer," breeding in balmy southern locations during harsh northern winters, and returning in the spring and summer to take advantage of vast foraging ranges. The strait is only two miles wide where the birds cross over to the Italian mainland and then spread out toward nesting sites in northern and eastern Europe.

We volunteered at an annual raptor-counting camp sponsored by the Mediterranean Association for Nature and the World Wildlife Fund. Then we crossed the island's north coast along the 38th parallel to Palermo, the capital of the island, where the Mafia is "kidnapping Sicily's water," according to one news headline (Colombo, 2004).

About forty-five volunteers help at Messina from mid-April to mid-May, during the peak raptor migratory season. The annual bird count has a dramatic history, with protective surveillance against poaching a key part of the effort. Anna Giordano is the spark plug who has fired up international volunteers here since 1984. That year, as only 3,198 raptors flew by, the team heard 1,185 gunshots. Birds were running a deadly gauntlet. At that time, thousands of migrating raptors were shot by local Sicilians during the annual migrations, though the birds had been legally protected since 1977. The senseless massacre was a traditional hunt for trophies; stuffed birds were displayed in homes and presented as gifts to show off shooting skills. A local folk legend even held that any man who could not kill at least one Honey Buzzard (hawks are called buzzards in Europe) during the spring migration would be cuckolded by his wife.

We made it to the camp, a Forest Service facility (38°16′N), as the sun set. It was windy and cold that night. Volunteer counters come and go each day during the counting season; ten were there when we arrived. Jean Paul Fiott from Malta had worked on the count for ten years. Michele Cento from Rome worked with the Italian League for Protection of Birds and was counting for his fourth year. It was the first birding adventure for Sophie Tyldesley, an 18-year-old English girl. In the prior three days the counting team had tallied 12,000 raptors, smashing the single-day count record for April, so spirits were high.

We spent the day on a windy 1,200-foot ridge in the Peloritani Mountains, with a view of snowy Mount Etna spewing ash in the distance. Groups of raptors kept arriving from the southwest, circling on thermals to gain altitude before attempting the water crossing. The ridge had previously been a hot spot for poaching.

As Anna Giordano told us her incredible story, her eyes constantly scanned for approaching birds. She usually picked out raptors before anyone else and quickly identified the species by the distant profile. Each time she would stop talking, confirm that the sighting was recorded by the team, and then take up exactly where her narrative had been interrupted.

"I was always in love with nature. When I was 15, we had a country house near Messina, and I heard about the migration of raptors in

Figure 16. International volunteers count migrating raptors at the Strait of Messina in Sicily and report illegal poaching.

the strait and that they were being shot by hundreds of poachers. I had known birds migrated at Gibraltar and the Bosporus Strait, but to think they were also by Messina, where I lived, no."

It seemed incredible to her that the birds were being shot, because hunting was prohibited in springtime. All raptors had been legally protected since 1977. She decided to see if what she had heard was true.

"Near here, I met a poacher who wanted to give me, as a present, three kestrels [small hawks] he shot. Of course I refused. I saw many birds killed, falling down into the bushes. I found shotgun shells with 'honey buzzards' printed on them in the local dialect. And then one day, I was surrounded by about thirty-five poachers. They became very aggressive . . ." She paused to look at approaching birds. "And everyone had pistols, guns, knives. They invited me to go with them to a 'meeting,' and said, 'What would your parents think if you do not come back home? Go! Do something else.'"

She answered a phone call from her office about more volunteers arriving that day. Hanging up, she said: "That was the first of their last days. I was very, very angry. I couldn't tolerate that they were shooting protected raptors from illegal hides [blinds] and nobody was doing anything. I was upset and angry at the same time and started this long campaign that has taken 29 years."

"You were only a teenager and yet your parents never discouraged you?" we asked.

"They never expected me to *not* do something, because they knew, if they said 'no,' I would do that thing." She laughed, recalling herself as a teenager. "They have the psychological approach to push me the right direction. Those first years, 1981, '82, and '83, I could do almost nothing out here. Then I started receiving great help from the Malta Ornithological Society and from Germany, and began this annual camp."

Another arriving bird flock interrupted her.

In the beginning, the volunteers stayed as close to the shooting as possible, in a concrete shooting hide, but after the BBC released a documentary film, *Anna and the Honey Buzzards*, poachers stopped letting them use that bunker, downhill from our ridgetop location.

"From 1985 to 1990, we used my mother's house for the camp, then we moved because she was tired of having forty persons in her house." Anna laughed, but her next words clarified the nature of the battle she had taken on: "No one would rent us a house. They were all afraid that some poacher could burn or do other damage to their houses."

That the Forestry Service had allowed the group to use the regional camp for the past two years signaled a long-sought change in agency attitudes.

"At first the government did not support us. They were thinking: 'Let them shoot. There are many other problems in the world.'" There was another pause for the approach of birds, this time Herring Gulls. "So in the beginning there was not too much help. When there was poaching, I would race all the way down the mountain to report to the Forestry or *carabinieri* [police], driving like crazy. Sometimes I would come into the office and the man was reading the newspaper, drinking his coffee, and he just told me that they don't have time to come. So I said, 'If you don't come immediately, I will write a charge that you are not doing your job.' So they began coming, but in the beginning, there were no arrests, because they didn't have the will. But they were coming."

The first arrest for poaching finally happened in 1984. "That day I was completely crazy, because they were shooting everywhere. I went for the *carabinieri*. We started . . . " She stopped and pointed. "Look there, a Honey Buzzard. Well, we started down the hill. The poacher saw us and started to run. So, then, I made a mistake; I shouted and ran after him. I found him hidden in the grass, but holding no gun and no birds, and when I turned my head I saw that I was alone with this man. The *carabinieri* had not followed. But I was not hurt, perhaps because he had hidden his gun, perhaps because he was an official, a traffic warden. Finally the *carabinieri* arrived, and a friend who was with me found the gun and dead Honey Buzzards."

"A scary moment!" we said.

"Yes! The man was arrested and began telling me, 'Oh, please, I am the father of a family. Please don't destroy me.' And I was thinking, yes, he is father of a family. Then I looked at the Honey Buzzards' black bills and said, "These were also the head of a family.'"

Anna was a young female challenging a male hunting tradition in a Cosa Nostra (Mafia)–dominated society. Her car was firebombed in 1986; fortunately she was not hurt. Her house was broken into. A dead falcon was left at one of the observation points with a warning note: "Your courage will cost you dearly." Only after shots were fired at her and some volunteers did the police change their do-nothing approach.

Gradually, local attitudes improved and her efforts began to pay off. Anna first felt things were changing for the better in 1989, when, without her prodding, *carabinieri* arrested a poacher for illegally shooting quail. The first year during which the counting team heard no gunshots was 2005. Shots are now a rarity, though toward the end of our day we heard two down in the valley as a flock of hawks circled above the distant farms. Anna and Jean Paul called the police and raced off, but they were unable to locate the shooter. When Anna returned, clearly angry, she declared, "Apparently 29 years is *still* not enough."

In 2006, something occurred that, Anna said, "gave me a sense of why I have devoted my life to birds. There was a man, one of the worst poachers and worst men I ever met in my life. He was always very aggressive. One day he stopped here for the first time, and he hugged me and said, 'Can we talk?'" They moved a little away from the counting group, and the man said, "I remember what you told us 25 years ago." And Anna thought, "My god, what did I say 25 years ago? I didn't remember." "You told us," the man said, "that one day we would start loving these birds and not shoot them anymore. I just came to tell you that you were right."

"I could not even find words," Anna said. "It gave me goose bumps. So I gave him some binoculars. Before, when he was going out from the farm, he would ask his son, 'Do you have the gun?' And now when they go out, he asks, 'Do you have the binoculars?'"

Anna's story reminded us of the 16-year battle to save Mono Lake, during which a few key people proved to be so important, though they never faced as violent a response as these anti-poachers did.

"I am convinced that each one can make the difference," she said. "At the same time, alone, you do nothing. I had only the responsibility to start the song that became a chorus."

"But somebody had to sing that first note," Janet pointed out.

Anna could not take such direct praise without making a self-deprecating joke: "Ah, I have a genetic deviation in my mind that makes me seek the justice that here does not exist. My life is dedicated to a fight, and any more is too much. In my next life, I want to be born with a passion for stamps." She laughed. "I am joking, but maybe I could be a judge so I could affect the people with power."

We asked Anna what she would most like us to write, what she most wanted to tell people.

"Never think that a fight cannot be won. Never, never."

More birds passed overhead to be identified and counted, and then she added: "Our role now is to prevent poaching. Before, it was to repress poaching; now to prevent. This Strait of Messina is a treasure of biodiversity. We have a high number of birds, fishes, insects, plants, sea mammals, and many endemic and rare species. If I become rich one day, I would like to make an international observatory; that has always been my dream. And a Secret Service for the environment," she added, laughing. "But I'm not able to find the funds. This is my fault. I can find a poacher, but I'm not able to go to a rich man and say, 'Oh, trust in us.'"

The BBC documentary *Anna and the Honey Buzzards* appeared in 1992, and she received the prestigious Goldman Environmental Prize in 1998, which annually honors grassroots environmental heroes from each continent. *Time* Magazine told Anna's story in 2000. We promised to also write about her effort, so that readers would help her dream for an international observatory come true.

By the time the 2010 count finished, 35,397 birds were tallied, including 31,124 Honey Buzzards, 1,769 Marsh Harriers, 583 Common Kestrels, and 520 Black Kites—a total of twenty-eight raptor and stork species.

Anna's courageous campaigning extends beyond the raptor count. She operates a wildlife rehabilitation center in Messina and, at the time we met, was fighting to stop construction of apartments and condominiums on unstable mountainsides. The Cosa Nostra seemed to be laundering money by building unsafe housing. In October 2009 thirty-seven people had died in mudslides. Because of her public opposition, she was having to fight eviction from the building that housed her rehabilitation center. She was also combating plans for a bridge over the Strait of Messina, a controversial project that many fear will only divert money to Mafia-controlled construction companies. But even if the bridge is begun and never completed, it could have a devastating environmental impact. The middle-aged dynamo's work is not yet over.

As we moved westward across Sicily on a very slow train, stopping at each local station, we passed through agricultural towns struggling with thirst. Although Palermo receives 29 inches of rainfall annually—10 inches more than San Francisco—and although the entire island receives 5.7 acre-feet of water as rain each year, "water trickles away, disappearing into the cracks created by poor management, corruption, and the Mafia. Lack of freshwater affects 53 percent of the population" (Colombo, 2004; an acre-foot covers one acre one foot deep). The Mafia's control of water dates back to the nineteenth century when *guardiani* (rural watchmen) charged consumers for water. Through the twentieth century, the island's water wells remained the property of prominent Mafia families.

Money from Rome and the European Union became available for constructing dams, reservoirs, and delivery systems, but they never materialized or were never completed. Some dams have been under construction for 20 years.

We saw some of the lush orchards irrigated by rich landowners near Palermo, a city with water problems despite a plentiful supply. Leaks in the public plumbing system have gone unrepaired (wastage estimates range from 27 percent to over 50 percent). Water theft by tapping into municipal pipes is a common practice. During a particularly severe summer drought in 2002, residents took to the streets to protest irregular water deliveries that sometimes left their household pipes dry for days. Analyzing that 2002 water crisis for signs that social unrest may have generated constructive change, researchers used academic language that avoided the word *Mafia* in concluding that "the balance of power . . . was skewed towards the 'traditional,' center-right political elites and power brokers . . . [who] proved extremely skillful in presenting their age-old solutions to the crisis, mainly based on the construction of additional hydraulic infrastructure" (Giglioli and Swyngedouw, 2008). The writers concluded that little had effectively changed.

Sicily's political culture has long grudgingly accepted this broken contract between society and its government with a philosophical shrug. We would like to think that Anna Giordano's successful battle for the raptors in Messina points toward a hope for broader change. It would be intriguing if water, that element essential for life, became another lever to further shift that island society toward a more productive pattern.

From Palermo, we caught a bus to Trapani, on the west coast of Sicily, and from there flew to Spain.

Spain's Coastal Lagoons, Water for Growth, and Iberian Lynx

The 38th parallel encounters the Spanish coast at Torrevieja (38°00'N), a town famed for its saltworks and singing competitions. There, at 00°42'W longitude, we were also west of the Greenwich Meridian (0° longitude) for the first time. Two immense saline lagoons cradle the town of Torrevieja, one colored green and the other pink. At La Mata, the green lagoon, we saw Brine Flies and bright-red Brine Shrimp (*Artemia salina*), Eared Grebes (*Podiceps nigricollis*), osprey, and avocets—species common to Mono Lake. But these lagoons were below sea level, the ocean was close by, and there were no snowy mountain peaks in the background.

The pink lagoon, colored by iron-fixing bacteria, is hypersaline. It tasted intensely salty, like the Great Salt Lake in Utah. Park ranger Antonio Saez explained that the two lagoons were natural features managed for centuries for salt production. A canal lets in the sea when the water level gets low. Machinery scoops salt off the lagoon bottom into huge drying piles. A colony of the rare Audouin Gull (*Gaviota audouin*) uses the saltworks as a nesting site.

Both lagoons attract wintering and nesting waterbirds, most notably more than 3,000 Eared Grebes, Northern Shovelers (*Anas clypeata*), and Red-Crested Pochards (*Netta rufina*), and up to 2,000 Greater Flamingos (*Phoenicopterus roseus*). Viewing blinds were spaced along a trail at the edge of the green lagoon, where a helpful English birder shared his spotting scope. (A Norwegian choir was singing in the picnic area, providing an unusual background sound for birding, but not an uncommon sight around Torrevieja.)

The largest lagoon on this coast is a few miles south of Torrevieja. The "small sea" of Mar Menor (37°49'N) provides more than 36,000 acres of bird habitat for nesting, migrating, and winter refuge. People can reach several swimming basins via a long boardwalk. Stretching out on our backs and floating buoyantly in the knee-deep saltwater reminded us of swims in Mono Lake, but in this experience, mud bathing was included. It was strangely enjoyable to spread over our bodies smooth black mud scooped from the bottom of the shallow lagoon.

This part of the Spanish Mediterranean coastline is called Costa Blanca. Most of the white, sandy beaches are backed by mile after mile of thirsty condominium resorts and golf courses. Southern Spain boasts more than 300 days of sunshine a year and draws sun seekers from northern Europe to Costa Blanca, Costa Brava, Costa del Sol, and even Costa del Golf. Outside the resorts, imported water irrigates flowers, fruits, vegetables, olives, and grapes that flourish in the warm weather.

Spain's water issues are almost identical to California's. There is a geographical mismatch between available water and demand. Long-distance aqueducts extend from the wetter north to serve urban and agricultural thirsts in the arid south. In 1973, the Tajo River was diverted from central Spain southeast across 175 miles to the Segura Basin in Murcia Province. In our drive to Mar Menor, we passed farm fields outside the city of Murcia and crossed a bridge over this Tajo-Segura aqueduct. The project was undertaken primarily to serve farm irrigation and stem groundwater depletion, just as California's Central Valley Project was meant to stop farms from overdrafting the southern San Joaquin Valley aquifers. But in both cases, irrigated acreage using the new water expanded, while aquifer problems persisted. Fertilizer and pesticide levels also increased in the Segura, and it gained the dubious distinction of being the most polluted Spanish river. In Spain as in California, water has also been shifted away from farms to serve the seemingly unending growth of tourism and urban development.

In 2004, when a new government took power in Spain, one of its first acts was to scrap another water transfer plan. Ebro River water was going to be taken from northeastern Spain, where there are 19,700 acres of natural wetlands and another 52,000 acres devoted to growing rice. The controversial project threatened the Ebro Delta environment, which had Special Protection Area status for birds, was part of the European Union's Natura 2000 Network, and was also a Spanish natural park.

The new government's alternative, Programa AGUA (Actuaciones para la Gestión y la Utilización del Agua), did not rely on long-distance water transfers. The focus became seawater desalination: twenty-four plants will generate 50 percent of the water in the AGUA Program. Another 20 percent is to come from wastewater recycling, 15 percent from modernizing irrigation systems, and the remaining 15 percent from other efficiency measures.

During our visit the largest desalination plant in Europe was under construction at Torrevieja near the east side of the pink lagoon, but work seemed stalled. Our requests to the construction company for a tour of the plant were never answered. The guard at the gate of the complex just shrugged and rolled his eyes when we asked when they might finish. The ranger's comment back at Mata y Torrevieja Park was, "*Muy* [very] *political*." A woman staffing the tourism visitor center told us that the project was controlled by the national government, so local people did not know much about it.

Developing sufficient renewable energy sources (the AGUA plan preference) to cover the cost of desalination and of pumping water inland, uphill, is a challenge that plagues ocean desalination efforts everywhere, as is mitigating the environmental damage to the ocean when concentrated brine is deposited. Europe's financial crisis was a complicating factor. One desalination plant was already operating near the Mar Menor, at San Pedro.

While planning the Spanish portion of our trip, we thought of Javier Grijalbo, an old friend from Janet's exchange-student days, 35 years earlier in Madrid. We reconnected with him by letter—a real one with a stamp and everything—explaining our project. Javi is a renowned illustrator and naturalist, the perfect guide. He and his wife, Virginia, traveled with us for five days through Andalucía, a place they know well, identifying every plant and bird (and all the good restaurants).

The Sierra Morena closely follows the 38th parallel across most of Spain between Andalucía and Andújar (38°07'N). Water drains down the mountains' southern slopes into the Guadalquivir River, which flows past the historic cities of Córdoba and Seville, then into a huge estuary, Las Marismas, before entering the Atlantic Ocean.

Much of the mountain landscape is *dehesa*—semiwild woodlands where olive trees and cork oaks are cultivated, but that are also managed for livestock grazing and hunting grounds. Black-Footed Pigs (*Pata negra*) root for acorns beneath the oak trees and provide the finest ham in Spain. The eastern half of the Sierra Morena also provides wildland refuge for the last stand of the Iberian Lynx (*Lynx pardina*), perhaps the most endangered wild cat in the world. Relatives of our North American bobcats, these lynx are smaller, standing about two feet tall, with tufted ears and sideburn whiskers framing their faces. Only 176 lived in the wild when we visited, with another 67 in a captive breeding program on the southwest coast inside Doñana National Park. With Javi and Virginia, we would go to both locations to learn about Spain's efforts to save the lynx.

North of Andújar, in the Jándula River Basin, the four of us spent a day with German Garrote Alonso, who works for the environmental and conservation arm of the regional government. When he spoke, Janet could immediately hear that he was from Madrid, his accent the one she best understands since her student days in that city.

The day before, a half-million people had swarmed into this region on an annual overnight pilgrimage to a hilltop shrine. German took us through locked gates into a 5,000-acre project area that is off-limits to the public. There, in the wildest landscape we would see in Spain, he works with wild Iberian Lynx and tries to expand the numbers of European Rabbits (*Oryctolagus cuniculus*), the critical food source for the lynx.

Rabbit disease epidemics are the major cause of the decline in the lynx population. Myxomatosis, a viral disease of South American origin, was deliberately introduced into Australia in 1950 and into France in 1952. It spread into Spain, quickly killing off 90 percent of the rabbit populations. As had happened in Australia, after an initial plummeting of numbers, the Spanish rabbit population showed signs of recovery as individuals with natural resistance to the virus increased. Then, in the 1980s, rabbit hemorrhagic disease arrived. The total reduction in rabbit numbers has exceeded 80 percent during the past 30 years. Road killings and hunting added to the problem. Rabbit hunting is one of the oldest traditions on the peninsula.

It has not proven easy to increase the rabbit populations everywhere, so an alternate tactic brings lynx to remote lands preserved for deer hunting, like the Sierra de Andújar, where rabbits still thrive. Of three pairs of lynx relocated during the winter of 2009–10, one female had given birth to three cubs, and another was pregnant when we visited. But some determined cats (their locations monitored with GPS) frustrated researchers by heading back to their original territories, traveling hundreds of miles.

After years studying photographs taken with motion-sensor cameras, German can individually recognize all of the lynx older than a year by their appearance. He puts out supplementary feed for the lynx once a week. He showed some of the cameras, strapped to trees, and several of his artificial rabbit warrens, built inside enclosures as an experiment to raise rabbit numbers.

Since the program began in 2003, numbers of lynx in the wild have increased by 40 percent, but it is still too early to be certain of success. As German said, "We have in our hands the power to restore the lynx. They are flexible and will respond. They just need to have the chance with protection of their habitat."

After we said goodbye to German, the four of us quietly waited and watched in the late afternoon near the Jándula River at a place outside the enclosed area where sightings are possible. We saw neither lynx nor rabbits. *"No importa, están allí,"* Javi later wrote in his blog: "No matter, they are there."

The following January, our naturalist friend went back to Sierra Morena and saw lynx four times. Through a spotting scope, Javi watched a lynx stalking, then pouncing and digging, rapidly throwing dirt. It paused to look around, then resumed clawing at the ground again until it pulled a small rabbit from its burrow, which it devoured.

Spanish people support the campaign to save the Iberian Lynx, though very few have seen them. Warning signs along the roadsides near Andújar prominently displayed a lynx face and the single word: *Recuerde*—Remember. At a store in the lynx region we noticed a decal with a painted lynx image and showed it to Javi. He did a double take. The artwork had been pirated from one of his beautifully crafted educational posters. He vowed to correct that misuse of the image on commercial products without his permission.

The Jándula River is one of many tributaries draining from the Sierra Morena toward the Guadalquivir, which becomes Spain's fourth-longest river and one of the country's largest as it passes the ancient city of Córdoba (37°53'N). Between 200 B.C. and 400 A.D., Rome controlled the city and made it the capital of Hispania Ulterior during the Roman Republic and of Hispania Baetica during the Roman Empire. Roman ships navigated from the Atlantic Ocean all the way upriver to Córdoba (boats today travel up only as far as Seville). The Romans built three aqueducts to deliver water from mountain streams into the growing city. Muslim Moors ousted the Romans in the eighth century A.D. and constructed the Mezquita, Córdoba's great mosque, on the site of a Visigothic Christian basilica. Under the Caliphate of Córdoba from 929 to 1031, the city became the most populous in the world (200 years before Merv topped that list) and was considered the intellectual, political, and economic center of Europe—an elevated status made possible by ingenious solutions to the water supply problem. Along the Guadalquivir Valley, during the al-Andalus period of Islamic dominance, thousands of waterwheels lifted river water into an extensive system of irrigation canals, similar to those used for the same purpose in Lanzhou, China.

Walking past gates on the city streets afforded glimpses of inner

Figure 17. A Roman bridge crosses the Guadalquivir River to the Mezquita mosque in the city of Cordoba.

patios, gardens, and fountains, and the lush use of water that typifies life in the city. The weather was sunny and hot, and it was a relief to return to the riverbanks, where new walkways had been completed since our last visit in 2003. Back then, we had noted a message painted on the wall above the riverbank: *Ríos Vivos: No Canelización* ("Living Rivers: No Channelization"), a sign of the challenges the watershed faced and the local population's concerns.

The route Javi and Virginia laid out for us passed through part of La Mancha Province and across Andalucía. Washington Irving, in his book *Tales of the Alhambra,* characterized the maturing of a girl to womanhood as "passing from the flat, bleak, uninteresting plain of La Mancha, to the voluptuous valleys and swelling hills of Andalusia" (Irving, 1994, 146). We looped back into the Sierra Morena north of Seville, which is designated as "natural park" land. Development is limited there, and the *dehesa* woodlands are protected, while the traditional agricultural economy and village life are also preserved. Pigs and cattle graze beneath cork oak and olive trees. We saw a Short-Toed Eagle (*Circaetus gallicus*) with striking yellow eyes, a single Black Stork (*Ciconia nigra*), and many White Storks (*Ciconia ciconia*).

Following small tributaries back to the Guadalquivir River, then downstream past the city of Seville, led us to Doñana National Park. Marshes at the estuary where the river spills into the ocean are renowned for waterbirds. The great river deposits sand and nutrients along the Atlantic Coast, creating extensive dunes and, on higher ground, habitat for scrub and *Pinus pinea* (pinyon pine). The region appeared unusually lush following an exceptionally wet winter in Spain, and those wetter conditions actually made it harder to see birds, since they could spread out across more of the verdant landscape. Javi and Virginia's great birding knowledge guided us to each habitat type, where they pointed out Flamingos, Spoonbills, Egrets, colorful Bee-Eaters, and animated English Chaffinches (*Fringilla coelebs*), very common there but new and fascinating to us.

Human activity at the edges of the park include charcoal production, beekeeping, wood gathering, fishing, and livestock grazing. Coastal condominiums and resort facilities lined the beaches, reminding us of what we had seen on the east coast of Spain. Much of the scrub habitat on private land has been converted to intensive irrigated agriculture (especially strawberries), causing groundwater levels to drop and interfering with the natural cycle of seasonal floods across the marshes. As in Turkey and Greece, the World Wildlife Fund has recently begun working with strawberry growers in Doñana to shift to drip irrigation, which uses half as much water. That is particularly significant because Spain grows 200,000 tons of strawberries each year by applying more than 16,000 acre-feet of irrigation water.

There were also threats from the upper watershed. When a wall at a toxic waste reservoir collapsed at a mine at Aznalcollar (37°31′N) in April 1998, millions of cubic yards of toxic mud and acidic water washed down tributaries of the Guadalquivir, contaminating farmland along the riverbanks and continuing to the main river channel. Fortunately, the plume stopped just outside the national park.

Doñana is the western habitat refuge for Iberian Lynx. We visited the Iberian Lynx Captive Breeding Center in Acebuche, inside the park, and met with director Antonio Rivas. While the in situ project we had explored in the Sierra de Andújar focuses on relocating wild lynx onto appropriate habitat, the ex situ breeding program at Doñana aims to build up captive numbers for future reintroduction. Nine cubs were born there in 2010. The breeding adults and cubs are absolutely off-limits to visitors. We were not allowed to let them see us or to see them, except on video screens, to ensure they remained leery of people. It was a privilege to see the action on video screens in a monitoring room staffed night and day. The ambitious effort is comparable to the successful captive breeding program for endangered California condors. The plan calls for developing a core of sixty to seventy thriving individuals before introducing adult lynx into three habitat zones: Andújar, the western Spanish province of Extremadura, and eastern Portugal. That number had actually been reached when we visited the center, though none of the captive-born cats had yet been released. Two females bred in captivity with a wild-born male did finally did make it out, in February 2011, to prowl for rabbits on the wildlands of Andújar.

During our days together in Spain, Javier's list of bird sightings in the Guadalquivir watershed grew past 120 species. After securing Javi and Virginia's promise to visit Mono Lake soon, we left them and traveled into the province of Extremadura, the "extremely hard" arid land of western Spain that produced so many of those extremely "hard" conquistadors who explored the New World. The city of Jerez de los Caballeros (38°19'N) was the birthplace of Hernando de Soto and Vasco Nuñez de Balboa. From there, we crossed the border into Portugal.

Portugal's Transported Town, a Solar Donkey, and the Azores

Outside the eastern Portugal town of Moura (38°08′N), where we appeared to be the only tourists, we visited the Alqueva *barragem* (reservoir) on the Guadiana River. The rolling hills were shaded by olive trees and carpeted with wildflowers of every color. The dam itself is not particularly large, but the reservoir behind it is the largest in Europe. The new reservoir, completed in 2002, filled for the first time during the winter of 2009, a wet season that followed years of drought. When we visited, no water deliveries had yet been made from the reservoir, because distribution canals were still being built. As with California's Central Valley Project and the Tajo-Segura system in Spain, the Alqueva water will go primarily to farms.

The small village of Luz was in the path of the rising water, so its three hundred residents were forced to move to a whole new town built at a higher elevation. Several displaced residents told us that the town provided by the government was just not the same, though the general layout was similar and their neighbors had not changed. They had watched bitterly as the old town and its surrounding forests were razed before the reservoir filled. A movie in the new visitor center brought tears to our eyes with stories of old fruit trees, favorite rosebushes, and gardens lost to the reservoir. The government's attempt to re-create the village as closely as possible, and its willingness to tell that poignant story in the film, rather than hush it up, contrasts favorably with the forced displacements of Kurds in eastern Turkey and the much larger relocations forced by Chinese dam building. Yet Luz remains a reality check on the human upheaval that often happens when waterways are dammed and residents uprooted.

"The day they closed the gates [on the dam], we brought a wooden 'white elephant' to the ceremony," José Martins said. It was a symbol of an unwanted gift that is difficult to dispose of and comes with upkeep costs far beyond its usefulness or value. José is the regional director for Quercus (Latin for "oak"), the major environmental NGO of Portugal. In his office in the city of Beja (38°01′N), he spoke quietly as he educated us about many issues surrounding the Alqueva reservoir.

Portugal and Spain had an agreement dating to 1968 to divide the Guadiana River water, but Spain pushed ahead with more than forty dams, large and small, mostly for hydropower, while Portugal waited until 1995 to start building its dam. The project was designed to store enough irrigation water for 250,000 acres. "But how are you going to have enough water for that, considering climate change?" José asked. "Here in the western Mediterranean there may be a lot less rain in the coming years. Also, we don't believe Spain will send us enough water in dry years to fill the reservoir, because Spain is using water on thousands and thousands of hectares of land. They have one dam almost as big as Alqueva."

Water quality was another concern. Plant roots separate irrigation water from the chloride and other dissolved salts it carries, leaving the salts behind, until soils become worthless for agriculture. Lost farming soils in the "Fertile Crescent" between the Tigris and Euphrates Rivers of the Middle East are a historic example; the salt problems on California's southern San Joaquin Valley farms are another.

"Another problem," José continued, "is that water is being transferred from one river basin into another. Most of the irrigated area is not in the Guadiana watershed, so less water is going down the river." That transfer alters chemical and biological factors, and no one knows yet what the effects will be on fish, algae, "or even bacteria—sometimes the little thing is the worst," he said. Water diversions will also reduce flows that transport nutrients and sand to the river delta, diminishing the life in that productive estuary. "But all this is taboo to speak about," according to José. "They say, 'Spain is doing it, building dams and transferring water between basins.'" The Portuguese government seemed determined to follow the example set by its larger neighbor.

In fact, the thirst for irrigation water was partly driven by Spanish corporations' recent purchase of olive-growing lands inside Portugal. "They have the money. Money!" José exclaimed. "It's just business. They don't want to know about the environment, about rivers, about riparian forests. They grow olives that now look like small vines. Intensive, superintensive. They kill everything else. That is an environmental disaster, in the last five or six years, on more than 120,000 acres here."

Hardy, drought-resistant olive trees evolved in the Mediterranean climate and were cultivated successfully by small-scale farmers for thousands of years. The complex ecological relationships of the olive-tree woodland habitat (known in Portugal as *montado*) were being lost to the new, intensive olive culture that forces higher yields with pesticides, herbicides, and irrigation.

The price for water in Portugal is set low for political reasons, despite a European Union directive that the price reflect actual cost. We explained how, in California, farmers get water at subsidized prices to help keep the price of food low for everyone, yet they can sometimes market that water to cities at a big profit. José was flabbergasted.

"That's completely . . . I don't know what to say. That's crazy! Well, here we have not gone that far." But the Portuguese have some "crazy" ideas of their own: "Some people believe, because we have the dam, it is going to rain a lot more here. Back in 1993 or '94, I remember, the president said, with Alqueva and more evaporation, maybe we can have more rain. I said, if that was right, would Spain need so many dams? Would places by the sea, like California, be so dry?"

The Quercus organization is only 25 years old. Portugal's long dictatorship had to end, José told us, before such an organization could exist. "Our mentality . . . " He paused, then said, "We don't have a tradition to join organizations, because of that political era."

José works on many issues, including nature protection, environmental education, energy, and climate change. Quercus has only a few thousand members and forty employees. Most financial support comes from corporate donations. The organization hosts a national television show every day and a radio show, *One Minute for the Earth*, that airs internationally. Quercus also owns and manages land for environmental education and a micro-reserve network to protect cork woods, orchids, and other species.

Toward the end of our conversation we discussed the world's largest solar electricity–generating facility, which we had seen near Moura. The array encompassed ninety-two individual panels in dozens of clusters. This was a central utility approach to power generation, though we agreed with José that it ultimately makes better sense for a nation to invest in solar panels installed on each house or business property. That day, an incongruous sight had appeared against the modern backdrop of the solar facility: a donkey cart driven by father and son with a load of hay.

"We have a good latitude for sun, yes?" José said, concluding our talk on an appropriate note.

On Portugal's Costa Azul, we stayed in the small fishing town of Porto Covo. Nearby in the port city of Sines (37°57'N) was the birthplace of Vasco da Gama, the seafarer who rounded Cape Horn and established the ocean route to India. We drove north up the Atlantic coast to Lagoa de Sancha, a lagoon entering the sea that attracts up to 13,000 flamingos and many storks. Signs similar to ones on the California coast announced the endangered status of Least Terns.

Returning to Sines, we located the fort beside Vasco da Gama's historic family home and his statue, staring westward into the Atlantic.

Figure 18. A donkey pulling a loaded cart, including its farmer owner and his son, passes one of the world's largest solar electricity–generating arrays near Moura, Portugal.

Nine hundred miles in that direction were the Portuguese Azores islands, our next destination.

The first thing we had to grasp in the Azores was the fact that São Miguel Island (the most populated in the Azorean archipelago) was *way* out in the Atlantic, 940 miles from the Portuguese mainland, yet still more than 2,000 miles from the east coast of the United States. All islanders must be self-sufficient, but the Azores are particularly distant from everywhere (37°46′N; 25°19′W).

Europeans knew of the Azores by the end of the fourteenth century. Medieval legends of Atlantis, a sunken land in the Atlantic, and of Sete Cidades, a fabulous kingdom of seven cities, became associated with the islands. Diogo de Silves officially "discovered" Santa Maria and São Miguel Islands about 1427. In the prior decade, Portugal's Prince Henry "the Navigator" had been operating a navigation school at Sagres,

Portugal, to promote the study of the sea and geography and to develop improved ships for long-distance exploration.

During Columbus's first voyage, in 1492, his flagship, *Santa Maria*, grounded on a reef in the New World and sank. The smaller ships, the *Niña* and *Pinta*, began their return trip to Spain in January 1493. A day after the two became separated during a storm, on February 15, Columbus reached the Azores island of Santa Maria. (Although it would make a nice story if the island name was tied to Columbus's first flagship, the island had had that name ever since a lookout on de Silves's ship exclaimed *"Santa Maria!"* when he saw the island on the horizon.) The *Pinta* missed the Azores entirely and so was not able to replenish its freshwater stores from the islands' springs and lakes; that boat made landfall on the Spanish mainland.

On São Miguel, we were never more than a few miles from a glimpse of the ocean, a constant reminder that this island is isolated in a vast sea. That close relationship is reflected in the name given to a North Atlantic feature that shapes the climate of Europe's Mediterranean region: the Azores High. Just as California's Mediterranean climate is controlled by a North Pacific High that blocks Arctic storms during the summer, then slides south to open the door to winter storms, the Azores High controls the weather patterns of the western Mediterranean and North Africa. Each summer, the subtropical high (or anticyclone) keeps the mainland dry and warm, and when the high shifts south in winter, wet northern systems may again deliver rain to the Iberian Peninsula.

Another constant on the Azores was evidence of volcanism. Lava rock is commonly used for building. Hot springs and fumaroles abound. In places, holes dug into the steaming ground served as cooking ovens. We visited steamy, bubbling Furnas, the thermal park of São Miguel, where a huge pool welcomed swimmers, and Fogo Lake, which occupies the crater of a caldera. After time in the warm pool, Janet's once white shirt will forever be a nice shade of orange because of the minerals. The nearby watershed includes Lombadas Springs, which supply most of the island's drinking water. Geothermal power is a key energy source on these isolated islands. We stopped at the geothermal plant that provides

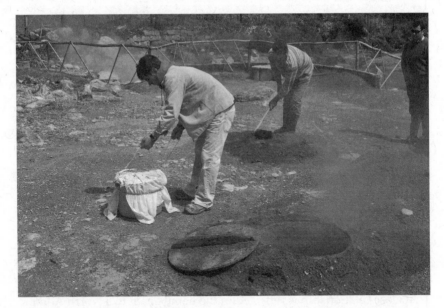

Figure 19.   On the volcanic Azores, islanders cook meals in holes dug in the steaming ground.

40 percent of the island's electricity, but as we walked through the open gate, a man ran up to shoo us out. He became much friendlier when we told him that the plant looked a lot like the geothermal facility outside Mammoth Lakes, not far from our Mono Lake home. He knew about that operation, yet photographs were strictly forbidden here as the design of the plant was proprietary information.

Sete Cidades only has one small village (not seven as the name of the legendary kingdom implies) in the center of a huge caldera that holds two lakes—one blue, the other green. A local legend tells of a princess and a shepherd, lovers who were not allowed to marry. The tears they cried in different colors gave each lake its distinctive hue.

According to a fisherman at Mosteiros, on the island's farthest western shore, fishing boats venture out only when the sea is calm—and one never knows when that may be. Natural bathing pools were protected from the waves by lava ridges along a rocky shore, and were used by local swimmers when the weather was warm, though not during our chilly mid-April visit. In fact a woman hiker told us that she had been warmer back home in Finland!

Some features of the archipelago reminded us of the Hawaiian Islands. The Azores had lots of flowers, particularly azaleas. The islands had wet and dry sides, determined by the prevailing winds, as in Hawaii, but this 38°N archipelago is definitely *not* at a tropical latitude (though pineapples grow there in greenhouses).

Every water issue, from pollution to supply, is exacerbated by population growth pressure. In most of the nations along our route through Asia and southern Europe, fertility rates have fallen close to or below the replacement rate (two children per woman). Globally, human population pressure is now a force mostly in nations south of the 38th parallel. Even where national population numbers are stable or declining, regional growth can still drive water issues. In 2012 the CIA *World Factbook* national fertility estimate for China was down to 1.55 children per woman. Intense population growth in and around Beijing, though, was behind the construction of aqueducts to bring water from the south. Fertility rates in Turkmenistan were 2.14, somewhat above the replacement level, and 2.13 in Turkey. Afghanistan and Tajikistan, two Asian countries we avoided because of safety concerns, had the highest population growth rates on the list of 38th-parallel nations, at 5.64 and 2.85, respectively. South Korea stands out from that Asian group, at 1.23, as does Japan at 1.39.

Across southern Europe, populations are either stable or in slight decline. Greece's fertility rate was 1.39; Italy's, 1.40; Spain's 1.48; and Portugal's 1.51. Despite the stable national population in Spain, thirst for development in southern provinces was another example of the persisting connection between water supply issues and growth. Sicily's water problems are not the same as mainland Italy's, while water for seasonal tourists is a particular problem for popular Greek islands.

Those nations that moved quickly to smaller families benefited from the "demographic bonus," in which growth in the number of young dependents who need nurturing and educating declines relative to the number of working adults. As Lester Brown noted in his book *Plan B*

4.0, this bonus allows productivity, savings and investment, and economic growth to increase. "Japan, which cut its population growth in half between 1951 and 1958, was one of the first countries to benefit from the demographic bonus. South Korea . . . followed, and more recently China. . . . This effect lasts for only a few decades, but it is usually enough to launch a country into the modern era. Indeed, except for a few oil-rich countries, no developing country has successfully modernized without slowing population growth" (Brown, 2009, 185). While modernization too often coincides with economic development that exacerbates water pollution and supply problems, ironically, it is also the only way most cultures can move beyond basic survival to address environmental concerns. While we traveled, the economic cycle swung downward, but the European Union upheld its laudable environmental goals and institutions, a reflection of that region's First World status.

During our stay on the Azores, we heard that Icelandic volcanic ash was disrupting air travel throughout northern Europe. We wondered how long we might have to remain on those islands in the middle of the Atlantic. The predicament brought home the real isolation of the Azores. Luckily, our flight route was far enough south to avoid ash, and we departed on schedule for the United States.

PART THREE

United States

Economists are winning; not hydrologists.

Mark Rude, on the overdrafting of the Ogallala Aquifer

Map 3. United States
1. Chesapeake Bay; 2. Charlottesville, Virginia; 3. Charleston, West Virginia;
4. Lexington, Kentucky; 5. Rolla, Missouri; 6. Dodge City; 7. Pueblo, Colorado;
8. Moab, Utah; 9. Pioche, Nevada; 10. Tonopah, Nevada; 11. Mono Lake;
12. San Francisco

Chesapeake Bay Watershed Education

Our east-to-west crossing of the United States began at the Atlantic Ocean on Assateague Island. We drove down the peninsula separating the east side of Chesapeake Bay from the Atlantic Ocean, known here as Delmarva, a blending of the names of the three states on the peninsula— Delaware, Maryland, and Virginia. Chesapeake Bay is 195 miles long and 35 miles at its widest; it makes San Francisco Bay look small by comparison. The watershed has fifty large rivers and thousands of smaller streams and extends north to Cooperstown, New York, and inland into West Virginia across 64,000 square miles. We were startled far inland by "Drains to the Bay" signs cautioning against allowing pollutants down storm drains—reminders that we were still within the watershed. We spent a week learning about the bay's beleaguered ecosystem along its intersection with the 38th parallel.

In 1608, after the infant colony of Jamestown survived its first winter on the bank of the James River, Captain John Smith and fourteen men in a small boat explored the bay for three months. They encountered native tribes, mapped the major arms of the bay, and left a journal record of the environment they saw. At that time, the water was clear enough to see through 30 feet to the bottom, where sea grass and dense reefs of oysters flourished. Both were keys to a bay ecosystem that has declined significantly in the past 400 years.

Today, over 16 million people live in the watershed. Pollution and overharvesting have decimated 99 percent of an oyster population that once filtered and cleaned all of the water in the bay every 3.3 days. Sedimentation smothered underwater grasses; only 70,000 acres remain where 400,000 used to grow. Algal blooms, fertilized by farm and sewage runoff, cause bacteria populations to increase as the algae dies and decomposes. That consumes oxygen and creates extensive "dead zones." Watershed forests and most of the bay-fringing wetlands that used to filter runoff and clean the bay are gone.

We passed through the village on Chincoteague Island and crossed a bridge to wild Assateague Island and the Chincoteague National Wildlife Refuge, at the edge of the Atlantic Ocean. A shell-covered beach stretches there for miles, home to endangered Piping Plovers (*Charadrius melodus*) and horseshoe crabs.

After refuge biologist Joelle Buffa explained her wildlife management programs, we climbed with Joelle and volunteer historian Myrna Cherrix to the top of the Assateague lighthouse—seven flights on a circular staircase. The Chesapeake is one of two large estuary bays (San Francisco being the other) that bracket the United States; each has a lighthouse positioned on the 38th parallel. Reaching this lighthouse required a walk through a forest thick with mosquitoes. Fortunately, Joelle provided jackets fitted with mesh and hoods to keep the mosquitoes from penetrating the fabric. We were casual, at first, about covering our faces with the netting, but learned fast.

The lighthouse was first lit on October 1, 1867, at the edge of the ocean. It towers above a forest now, because sand deposits built up new land around it. The south end of the island keeps growing, though the entire low-lying island may be covered by the rising sea in the following century, according to climate-change projections. Meanwhile, the light still burns brightly at night.

Within the wildlife refuge, there were many familiar birds: herons, egrets, geese, and Mallards, but a Tricolored Egret (*Egretta tricolor*) and Laughing Gulls (*Larus atricilla*) were new to us. The Chincoteague pony is the most famous wild animal in this area.

Local legend claims that ponies swam onto the island after a Spanish shipwreck. It is more likely that settlers first put them there to graze where no fences were needed to contain them. Every July, the ponies are gathered to swim between the two islands, so that yearlings can be sold—a way to control the island population (which has no predators) while raising money for the Chincoteague Volunteer Fire Department. The children's novel *Misty of Chincoteague*, by Marguerite Henry, was inspired by a true-life story of children determined to buy a pony at the annual auction.

The north half of Assateague Island is a national seashore operated by the National Park Service. There, the pony herd is not culled by annual drives and auctions, but population size is managed by injecting mares each year with contraceptives.

Joelle invited us to spend the night at her nearby home, where our GPS confirmed that she and her husband, Clyde, live exactly on our latitude line. Clyde retired after many years directing the Don Edwards National Wildlife Refuge in San Francisco Bay, and Joelle had previously been in charge of the Farallones Islands, off the coast of California, so their move east had essentially followed the 38th parallel.

We joined two school classes on environmental education boat trips that started from different points on Chesapeake Bay. The first trip, launched from the Delmarva peninsula, required a ferry to Smith Island (38°00′N), 10 miles out into the bay, where the Chesapeake Bay Foundation operates an education center. For much of the trip we could not see land. The bay felt like a huge calm sea.

Environmental education coordinator Krispen Parke hustled us directly over to another boat, one of the foundation's fleet, loaded with students from Clarke County High School in Berryville, Virginia, west of Washington, D.C. They were all from rural farms and "knew all about plows, cows, and sows," their teacher said, but this was their first experience on the bay. Their goal was to better understand how agricultural practices in the upper watershed affect nutrient and sediment deposition and, ultimately, the whole bay ecosystem. The program is the reverse of a strategy that brings school kids from Los Angeles to the Mono Lake Basin to help them understand how their actions, downstream in the distant city, affect their upper watershed. One of the Chesapeake Bay Foundation's guiding philosophies is that "the place to teach people about the bay is on it and in it."

Off we went, back onto the water to learn about its living creatures. Chesapeake's average depth is only 22 feet, but during that cruise we were usually in less than 5 feet of water. Even so, the bottom that was so clear in Captain John Smith's day was invisible to us because of the murkiness. The biggest problems in the bay are nutrient deposition from agriculture and urban runoff, and the sedimentation that smothers the bottom grass.

"The oyster is a good example of how we sometimes underestimate the importance of creatures by considering them mainly as they relate to our bellies and our commerce," local author Tom Horton has written (1991, 27). Yet foundation educators know that lessons are often best remembered when related to other senses. That is why they had the students pull up seventeen crab pots, put out the day before, and collect dozens of soft-shelled blue crabs that would become their dinner that night. The students measured each crab's shell and learned to tell males and females apart. Interestingly, there are no size limits on females, since their size does not relate to age. Oysters were put back into the water because the harvesting season had not yet started. An assortment of other bottom dwellers also showed up, including pipefish, grass shrimp, jellyfish, and bits of the grass that supports much of the bottom life.

The students had placed their crab pots along a state boundary. Krispen told the story of Ben Marsh, an oysterman and storekeeper who successfully lobbied to have the Virginia-Maryland border placed right through the Smith Island archipelago, so that fishing could occur in the waters of whichever state had the most favorable regulations in a given season.

On the ferry returning to the mainland, we met Tom Horton, whose books we had read about the Chesapeake Bay Foundation's efforts and about raising his family on Smith Island. Tom has not lived there since the 1990s, but was present that day with another school group. Protecting Chesapeake Bay is an overwhelming task, but, Tom said, "it would be a lot worse by now if we hadn't done anything."

He was referring to the foundation, which had been established 45 years before. "Save the Bay" bumper stickers are familiar across the watershed. (The same communication medium also served the campaign to "Save Mono Lake" across California.) With more than 200,000 members and 160 staff, the organization can now claim considerable success; yet in 2010 its annual "State of the Bay" index, which monitors thirteen key ecosystem indicators, stood at only 31 (a perfect score would be 100).

Our second day with a school group on Chesapeake Bay was on Captain Jack Russell's skipjack boat, another foundation vessel that works out of St. George Island (38°09'N), where the Potomac River enters the west side of the bay. We joined third-graders from Beach School in Chesapeake Beach, Maryland, on a cruise that was an interesting contrast to the day spent with high school students.

Skipjacks are rakish sailing vessels that were used to work oyster beds before Maryland laws allowed power-driven dredges. The activities on Jack's boat were familiar, though they *sounded* different on this side of the bay. We "drudged" for "arsters," in St. George Islanders' dialect, which are called "ersters" over on Smith Island and the eastern shore, and "awsters" up in New England. Jack had shucked oysters for 30 years, but there had been no commercial product taken since 2000. He has shared his knowledge with more than 60,000 school kids.

Cap'n Jack's approach was perfectly adapted to third-graders. Having conducted many school groups on tours at Mono Lake, we enjoyed watching a professional at work. Before boarding, he had all the kids stand in a ring holding a rope, called a "line" by sailors. When he told them, "You may get fish guts on you today," the kids cheered loudly (we wondered how that line would have worked with the high school crowd). Jack said, "Kids, y'all have to do *all* the work." That produced silence.

Figure 20.   Captain Jack Russell shows a schoolgirl how to hold a crab without getting pinched. Her class is on a cruise to learn about Chesapeake Bay.

"Parents don't get to do any." The kids yelled, "YES!" falling for the old Tom Sawyer whitewash-the-fence trick. As life preservers were passed out, each inscribed with the name of some part of the boat, the teacher had to raise her voice: "Girls, it doesn't matter what color they are!"

Before launching, we looked at a live eel and a Bardog in tubs. Jack explained that Bardogs, also called Toadfish (*Opsanus tau*), have no commercial value, yet were in decline—an indication of problems in the bay ecosystem that go beyond fishing pressure. One mother, along as a chaperone, missed Jack's point and asked, "Isn't that good?" (apparently because the Bardog is commonly called a "trash fish"). Jack's simple answer: "Everything has its place."

When he asked the students what they thought they might see that day, one answered, "Goldfish?" "Well, OK," Jack said. "Some people do throw theirs away into the bay." We appreciated how he never discouraged participation by reacting negatively to "wrong" answers.

Ospreys nested on every channel marker post as we motored out. Jack kept talking, teaching us names for the parts of the boat, making sure to mention the name on each student's life jacket. He taught everyone how to tie a bowline knot.

As nets from the skipjack dragged the bottom, Jack explained the sorry history of overfishing and ecosystem decline that was now familiar to us. Oysters each filter 50 gallons of water a day and used to clean the bay's water, but numbers had slid downward ever since the 1970s. "There are too many people. Their wastes have to go somewhere, and that's right here," Jack said.

A few seconds after bringing the dredge off the bottom, Jack pried open an oyster and turned to David to show the kids how to eat it. They watched closely as it slid on down.

The Cap'n told us that oyster season was in the months with an *r*: September through April. Then he recited part of Oscar Nash's poem "Oysters" (1931, 30):

> I'd like to be
> An oyster, say,
> In August, June,
> July, or May.

We pulled up crab pots, and Jack taught the kids how to hold a crab to throw it back in the water without getting pinched by its claws. Blue Crab numbers were the only bright spot on the Chesapeake, Jack said, "because crabs like trash and polluted water." While baiting the empty crab pots, Jack had the kids gather around for a solemn ceremony thanking the Alewife baitfish.

After snacks were handed out, one boy asked the teacher, "Do we get recess now?" She told him, "This day *is* your recess."

The overall message Jack wants to send away with the students is: "Place a piece of the bay in your hands. Maybe you will be impressed enough with what you see to take better care of it than my generation did. It's our bay . . . let's pass it on!"

One of the student letters posted on the wall in Jack's office reads: "I learned the left side of the boat is called port and the right side is called starburt [*sic*]. My favorite part was getting my hair wet. I loved the trip. Jenna." She drew a picture of the boat on the water, with the sun overhead wearing sunglasses.

The Rappahannock River
and Mattawoman Creek

To reach the west side of Chesapeake Bay, we traveled south under its 18-mile mouth through a tunnel. Joelle, the biologist at Chincoteague, had suggested we visit a hawk-surveying and songbird-banding site along that route, near the end of the peninsula. Just as the Strait of Messina in Sicily funnels migratory birds, the Delmarva peninsula leads raptors, songbirds, shorebirds, some waterfowl, and even butterflies toward the bay's mouth on their seasonal migrations. The Coastal Virginia Wildlife Observatory staff and volunteers have counted birds there since 1963.

On a calm, sunny morning at Kiptopeke State Park, near Cape Charles, we walked from our campsite to the hawk-watching platform. An average of thirty hawks, eagles, and falcons passes overhead every hour during the migration. Coopers and Sharp-Shinned Hawks seemed in a rush to make the long crossing as they hurtled on the wind straight south, rather than soaring in circles as in Sicily. There were lots of Yellow-Shafted Flickers, too; the group had counted more than four hundred already that morning.

At a bird-banding station in the nearby woods, Ann Gilmore and several volunteers gathered songbirds caught in mist nets to measure and band them. Janet got to release a Northern Pirula back to the wild with a tiny new band on its leg, and wished it well on its long trip to South America. It was hard to believe such tiny bundles of fluff, feather, and feistiness can fly so far. Another treat was handling a cardinal, quite common here, but strikingly red to us westerners.

After we made our own crossing, we stopped briefly at George Washington's Birthplace National Monument, along the bank of the Potomac (38°12′N), where a bald eagle flew by. That seemed an appropriate and auspicious sighting.

In Fredericksburg, Virginia, we walked through an idyllic woodland with John Tippett, the slim, articulate director of the Friends of the Rappahannock River, to see the former site of Embrey Dam (38°19′N). After many years of effort by John's group to have it destroyed, in 2004, Navy SEALS blew it up, allowing the river to again flow freely. Having filled with sediment, the dam had lost its usefulness for hydroelectric generation, and alternative ways to capture water upstream had been developed to supply Fredericksburg. John brought along a photograph taken, from where we stood, of the free-flowing river and overlaid with a ghost image of the dam at the instant it exploded. He told us how exciting it was that day to set the river free. "The fireworks of a spectacular explosion is not the norm in environmental protection," John said, "which usually consists of lobbying, action alerts, and fundraisers."

A nearby plaque commemorated the achievement: "Since the mid 19th century, the Rappahannock river had been holding its breath behind a wall of iron, concrete and wood. The rapids formed where the Embrey dam once stood are now and forever to be known as John W. Warner rapids," in honor of the United States senator who was a leader in the campaign. That positive story brought home to us, again, how often a key person at a certain moment in time (in this case, Senator Warner) can make a difference for a special place.

Until the dam was removed, anadromous fish including herring, American Shad (a threatened species), Alewife, and Striped Bass could not complete their life cycles by returning from the bay to spawning grounds upriver. Through the twentieth century, fish returning each spawning season swarmed in frustrated masses at the base of the dam. As part of the Embrey Dam campaign, each autumn, on the third Saturday in September, employees of the Game and Fisheries Department had electronically stunned the fish long enough for volunteers to scoop them into buckets. People formed a human chain to move the fish and release them above the dam. After those fish spawned, their small offspring could wash over the spillway to reach the bay.

That effort, mostly symbolic, kept people involved in the years before their goal of dam removal was achieved. It reminded us of the annual Mono Lake bucket walks that delivered water from above the creek diversion dam down to the lake and helped inspire a generation of activists. The Friends of the Rappahannock still hold a festival on the third Saturday of September.

The water-supply alternative to the big dam involves a totally different approach. Today, many small dams provide storage in side drainages filled by pumping water from the main river channel. Strict regulations ensure that the pumps "only sip from the river when it's running at high flows," John said, to avoid affecting the river ecosystem.

The Friends group also helped secure city ownership of the riverfront along the entire Rappahannock watershed and 30 miles of permanent conservation easements. "That took three years, two city councils, partnerships with The Nature Conservancy and others," John said. An endowment also funds a "river cop" to enforce regulations along the river.

As in many places in the vast Chesapeake Bay watershed, the group's long-term battle is with sprawling development. Wherever paving and construction make land impervious to water, nutrient and sediment runoff increases. Chesapeake Bay Foundation scientists have documented a threshold—10 percent coverage of the land—beyond which drastic declines in biological health occur in the bay. "We've made progress on all of the major point sources, like wastewater discharges and agricultural waste, but have been going backwards when it comes to urban runoff," John told us. "In fact, it's counteracting everything else." Although a billion dollars has been spent to upgrade community wastewater treatment plants in Virginia, what runs off yards and parking lots is equally important because of population growth and unchecked development.

We crossed the Potomac, leaving Virginia to reach Maryland's Mattawoman Creek, one of the most pristine streams draining into Chesapeake Bay, despite its location just 25 miles south of downtown Washington, D.C. In 2009 the creek was number four on the American Rivers Organization's annual listing of the most threatened U.S. rivers (the Sacramento–San Joaquin River system, in California, led that year's list). It was no surprise, by then, to learn that the biggest threat to the Mattawoman watershed was sprawling development. Things were at a crisis point because a major cross-county highway proposed in the upper watershed would harden many miles of land and serve even more suburban sprawl.

Smallwood State Park campground is beside Mattawoman Creek (38°33′N). The "creek" name had us expecting a much smaller river. In Mattawoman's wide channel, freshwater is pushed back up the creek by each rising tide that enters from the bay. The campground was almost empty. We fell asleep listening to a Great Blue Heron's croaking call and geese honking overhead.

The morning dawned overcast but calm, and we met Jim Long, president of the Mattawoman Watershed Society, and Bonnie Bick, vice president of the society and the Sierra Club representative on the Mattawoman issue. The four of us kayaked up the creek, gliding among huge round pads of American lotus while Great Egrets winged overhead.

Proximity to the national capital is the root of the Mattawoman's problems. The watershed of the seven-mile-long river contains the fastest-growing region in Maryland and, when we were there, had already reached the tipping point at which more than 10 percent of land was impervious to water. The paving, concrete, and construction that would follow in the wake of the proposed new highway could ultimately alter runoff on 25 percent of the creek's watershed. Nearly two-thirds of that new road was finished, leaving only the stretch closest to the Mattawoman. Permits are required to disturb wetlands, and Jim's organization was hopeful that national and state agencies would deny them.

Jim, a bespectacled government physicist, told us what moved him to work so hard to protect the Mattawoman: "I started kayaking all around the bay and eventually found Mattawoman," he said. "I realized how rare it is, and it became my biological center."

On our return to the launch point, the board of directors for the American Rivers Organization had a large group ready to begin kayaking. Jim and Bonnie were to attend the board's dinner that night, seeking national support for their local efforts.

Protecting the Chesapeake Bay is a daunting challenge because runoff problems originate across such an enormous watershed. "We need to learn how to behave, and quickly," author Tom Horton wrote in 1991. "We have already ensured that nearly a generation will grow up in parts of the watershed without the ability to catch shad and rockfish; ensured that a generation or two will never wade as children in grassy, clear shallows in pursuit of softshell crabs; and risked that many generations to come will take congested highways and strip development and sprawling suburbia as their normal environment" (Horton, 1991, 288).

In 2010 the Environmental Protection Agency created new Total Maximum Daily Load objectives for the Rappahannock and the main stem of Chesapeake Bay that aim for a 25 percent cut in nitrogen and phosphorus and at least a 16 percent reduction in sediment. The states surrounding the bay must develop plans within the next 15 years to undertake specific two-year actions aimed at pollution reduction. Also, President Barack Obama issued an executive order in 2010 that declared Chesapeake Bay a "National Treasure" and called for a strategy to restore and protect the bay, its watershed, and its resources.

In November 2011, two years after our visit, the Mattawoman Watershed Society celebrated news that the Maryland Department of the Environment had denied the cross-county connector highway project's wetland permit for failing to properly address its environmental impact.

In one day's travel the 38th parallel next took us by President James Madison's Montpelier home, James Monroe's birthplace on the Potomac River in Fredericksburg, Thomas Jefferson's famous Monticello home in Charlottesville, and, in Staunton, Virginia, to the birthplace of Woodrow Wilson. Jefferson was taken with the idea of growing grapes and olives at Monticello after visiting Italy, thinking that they might thrive there because of the shared latitude (38°01'N). In the seventeenth century, the Jamestown settlers tried to grow orange trees, because they had originated on the same latitude in Spain. Both experiments failed, as latitude is not the only factor controlling climate. A guide told us that Jefferson, who kept exact weather reports and was an inventor and amateur scientist, "was most likely very aware of his location on the 38th parallel latitude; it would be very Jeffersonian."

Atop the Blue Ridge Parkway, we stopped to look back to the east on an exceptionally clear day. The next high point along the latitude back that direction was a peak on the Azores Islands. We left crabs, creeks, coasts, and curlews behind and headed west, where the continental United States awaited. Scarlet fall color added a touch of flame that grew more intense as we crossed the West Virginia state line (38°07'N).

Mountaintop Removal Coal Mines
in West Virginia

Pocahontas County, West Virginia, straddles one of the continent's water divides and is known as the "Birthplace of Eight Rivers," streams that send water either back down to the Chesapeake Bay or westward to the Ohio River, part of the vast Mississippi River watershed that drains the middle of the continent. At a festival in Hillsboro, West Virginia, at the historic birth home of author Pearl S. Buck, several volunteers enthusiastically named the eight rivers for us: the Greenbrier, Elk, Williams, Cherry, Cranberry, North Fork of the Cheat, Tygart, and Gauley.

That same day found us at the Cranberry Glades in Monongahela National Forest (38°12'N). The glades are strangely fascinating because Arctic bog species are found there, delivered long ago by an ice age glacier. Cold, snowy conditions at 3,366 feet above sea level, plus local acidic soils allow the plant community to persist far outside the Arctic.

The brown, green, and golden hues of pitcher plants, cranberry moss, spruce, hemlock, and alder trees enticed us to walk twice around the loop trail. If Thomas Jefferson and the Jamestown settlers had visited the Cranberry Glades, they would have better understood that latitude is not the sole factor dictating plant growth: microclimates, topography, and soil types are even more important.

A raft trip on the lower Gauley River was an exciting 14-mile white-water experience near Fayetteville, West Virginia (38°04'N). Whitewater rafting has become an important alternative to the coal industry that dominates the regional economy. Some of the most popular raft-ing stretches in the eastern United States are along New River Gorge National River (one of only four "national rivers" in the National Park System), and along 40 miles of tributaries in the Gauley River National Recreation Area, where ten local guide companies serve about two hun-dred thousand people each year.

Our raft guide was Justin Reynolds, the son of a coal miner, who had run the New River two hundred times by the time he was 16 years old. A dam on the Gauley provides steady water flows for barges head-ing downriver toward the state capital, Charleston, except when pulses are released to serve recreational rafting. Trips on the upper and lower Gauley are timed to ride the "bubbles" of water released from the dam on six weekends each fall. Our arrival during the short rafting season was serendipitous.

It was a wild ride with Justin, who sought out "surf holes," where the boat sat in one spot while the river crashed around us. When our raft sprang a leak, Justin had his hands full trying to steer as the boat gradu-ally deflated and our anxiety inflated. Luckily, the river was busy with other rafters who loaned us a pump.

The owner of our raft company, Skip Heater, grew up locally, and he shared his knowledge about regional history over breakfast. A key rea-son for establishing the New River Gorge National River in 1978 was to stop a coal strip mine planned for the gorge. After the Civil War, Skip told us, "carpetbaggers" bought land for coal mining. In 1873, railroad tracks were extended along the river to serve the mines. Today, 70 per-cent of the natural resources of West Virginia are owned by outside corporations. "We were sold out ages ago, and they're bleeding the area dry," Skip said. "The parts of West Virginia that today have poverty problems are the coal areas. Thriving, clean communities are not where there is coal."

The region had a long history of labor-management conflicts between miners and their boss companies. Skip told us about America's worst industrial disaster, which "almost no one knows about." Drilling of the 3.7-mile Hawks Nest Tunnel began in 1929 in order to move water from behind a dam on the New River to a hydroelectric plant at the Gauley Junction. The project took three years. Miners for Union Carbide drilled through sandstone, but were given no dust control or breathing protection. Deadly silicosis killed thousands of them. The fate of those Depression-era workers, desperate for work despite the deadly conditions, remained nearly invisible to most of the world. Many who died were buried in unmarked mass graves. A historical novel, *The Hawk's Nest*, by Hubert Skidmore, was published about the disaster in 1941, but the Union Carbide Company forced Doubleday to pull all copies from publication, helping to bury the grim story.

In the past two decades, mountaintop removal (MTR) has become the preferred coal-mining technique. A highly mechanized approach that employs fewer miners, it has allowed companies to reduce the number of coal industry employees in West Virginia from 150,000 in 1950 to about 18,000 today. In the coming days, we were to learn much more about MTR coal mining's devastating impact on rivers, soils, and vegetation; the threats it poses to human health and property; and the political powers in West Virginia that defend the controversial practice, choosing short-term economic incentives over long-term societal and environmental health.

Driving from Charleston, we passed a billboard that read "West Virginia Coal, the Real Power behind America." We were about to meet with Larry Gibson and learn how powerful King Coal really is in the state. For 250 years, Larry's family has owned the summit of Kayford Mountain (37°58'N). Known as the "Mountain Keeper," he is a staunch protector of his home and opponent of MTR coal mining, which turns mountains inside out as coal companies blast them apart to get at thin seams of coal. The coal is separated out, and the rest of the mountaintop, now considered "debris," is shoved downhill into stream canyons. More than five hundred mountaintops totaling 1.2 million acres have become flat terraces covered by unproductive rubble. During the second Bush administration, waivers to Clean Water Act restrictions were issued that redefined mining debris as acceptable "fill material" in waterways. The requirement to maintain stream buffer zones under the Surface Mining Control and Reclamation Act was also lifted. As a result, more than 2,000 miles of streams have been buried.

We drove up a two-lane highway through a canyon where every few minutes a large truck full of coal passed us headed downhill. A small stream paralleling the road was red with acidic mine drainage. Larry's property on Kayford Mountain is a leafy island in a sea of mining operations. He made us welcome, but only after a wary greeting. His resistance had generated drive-by shootings and several violent incidents. He had a bulletproof vest hanging on a wall in his cabin, a gift from some supporters. He had never worn it, but kept it as a reminder to be cautious around strangers. He told us that the truck drivers had surely noticed and reported, on citizens band radios, our arrival that day in a Prius (not a common car in West Virginia).

In 1993 the Massey Coal Company made Larry and his family an offer for their land. "I got them angry when I wouldn't let them have it," he told us. "They said, 'We're going to take you off that mountain, give you some buying power. We're going to give you $140,000. How does that sound?' Like I should be flattered. They didn't understand that there's some things in this life . . . " He paused. "What do you have so close to your home circle of life, that's not for sale? How far are you going to go to stand for it and what would you do?"

Larry was 63 years old when we visited, a short, spry, muscular man with a white mustache. Although we were eager to see the MTR site that would become visible after a short walk through the trees, Larry first invited us into his simple cabin and made coffee. "My wife says I should serve toothpicks with my coffee," he joked, but it tasted fine.

Larry asked us to look at his walls, where he had framed copies of magazine articles and photographs of celebrity visitors to his hilltop. "Notice how everything's on with screws? When they're mining, dynamite goes off ten to fifteen times a day." Today, Larry must haul water to the cabin because blasting lowered the water table. Twenty-five tons of explosives are used each day across Appalachia.

MTR destroys not only natural resources but also a way of life. At one time, one hundred thousand people lived in the canyon below Kayford Mountain, called Cabin Creek Hollow. As the mines bought people out, "they knocked over 20-something schools," Larry said. "We got one left. We had a school here, on top, and a church, a general store, about sixty families living right here. This land was everything. It was our doctor. Our food source. Our water source. Our medicine, shelter, warmth. Most people don't understand when I say things like 'A mountain is a live vessel. A live body.'"

The family moved off the mountain when Larry was 11. After he retired from General Motors in 1985, Larry moved back to the mountain. "It took me four years to clear plants that had covered the family cemetery," he said. "We had two family cemeteries, one on this side of the mountain and the other surrounded by coal. That one they destroyed."

The federal acts that regulate mining and protect water could have been enough to stop MTR valley fills if exemptions had not been issued, but the problems go further. Toxic chemicals leach downstream, and the thousands of gallons of water used to wash each ton of coal during processing leave a toxic slurry that must be stored behind dams. Sludge reservoirs have become a looming threat to communities in the hollows downstream.

"We've got young kids going to school" Larry said, "beneath a preparation plant less than five miles from here as the crow flies, at Marsh Fork School. You've got one of the largest reservoirs in the world, called Shumate—9.7 billion gallons. That dam is 89 feet taller than the one on the New River Gorge. It's not completely full yet. If it breaks, 17 miles away from the breaking point, the line of waste coming at the people will be 40 feet high. How many people you think will be dead by that time? We have twenty-five of these storage dams in this area."

The concern is very real. On February 26, 1972, a flood of waste sludge cresting over 30 feet high descended on sixteen towns in Buffalo Creek Hollow (37°46′N). Within a few minutes, 125 people were dead, 1,121 were injured, and more than 4,000 were left homeless; 561 residences and 30 businesses were destroyed. The Pittston Coal Company, trying to avoid liability, called the dam collapse "an act of God." We would soon visit a town in eastern Kentucky hit by a coal-waste flood in October 2000 that swept down 90 miles of river.

It seemed inexplicable that, with so few people actually employed in coal mining today, the majority—those who had lost jobs and property—would not have a stronger voice in the region's politics, which currently favors the industry.

"For me," Larry said, "I've been fighting this so long that I think everybody has the sentiments that I have, but I've seen that people are intimidated. That's why the coal companies are able to do what they're doing. You make reference to democracy, something we're supposed to have. In this part of the country, we don't have it."

Walking a leafy path through an orchard to the end of the ridge, we realized there were no birds flying through the branches. "From 149 species of birds," Larry said, "we're down to 39. Every year I have people come in, checking our birds for us. I don't hear any birds."

At the crest of the hill, suddenly everything changed. As far as the eye could see, the mountains were gone, peeled away on three sides of Larry's land. We looked down on flattened terraces of rubble. At the distant edges, valley-fill debris created huge walls where streams used to run. Larry had tears in his eyes as he surveyed the scene, and so did we. The damage was so permanent and so short-sighted. We thought about how we would feel if our hometown mountains were blasted into rubble.

Out in the open, away from the trees, the wind suddenly began gusting. The ground at our feet was rough cobbles, some baseball-sized. "It won't be the same [again] in my lifetime, maybe forever," Larry said, kicking at the rubble in frustration. "What I'm showing you is what it will look like for many centuries to come. It takes about a thousand years for an inch of topsoil to form in these mountains." Green-tinted hydroseeding helped provide a temporary illusion of green. "That's their 'restoration,'" Larry told us, "with an introduced Asian grass."

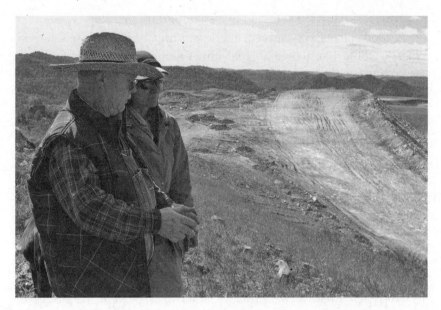

Figure 21.   Mountaintop removal coal mining has chewed away Kayford Mountain, to the edge of "Keeper of the Mountain" Larry Gibson's family property.

The total MTR acreage spreads across four states. As of 2008, West Virginia had 136 mountaintops removed totaling 352,000 acres; 6 mountains in Tennessee totaled 78,000 acres; Virginia's numbers were 156,000 acres on 64 mountains; and Kentucky was most affected, with 574,000 acres on 294 mountains (Geredieu, 2009).

When we visited Larry, the Environmental Protection Agency (EPA) had just decided to re-evaluate all mountaintop removal mining permits and to assess effects of MTR valley fills on aquatic ecosystems. The draft report that the EPA produced concluded that streams are being permanently lost due to burial, chemical pollution is occurring downstream (including selenium concentrations elevated to levels toxic for fish and birds), and aquatic communities "are consistently and significantly degraded" (U.S. Environmental Protection Agency, 2009). Before the EPA made any findings public, West Virginia politicians derided the decision to evaluate the effects of MTR as, in itself, a threat to jobholders and the state's economy.

To us, re-evaluation sounded like an important first step.

"A good sign," Larry agreed, but he was impatient with the slow process. "I would like to see all coal mines stopped today and see them leave. If they left today, they wouldn't destroy anything else. I call for the abolishing, the abandonment of coal, because it kills. Simple. It kills."

What we saw happening on Larry's mountain reminded us of techniques used in modern gold mining that chew up hills, heap the material produced, and leach out the gold with cyanide. It also reminded us of late-nineteenth-century hydraulic gold mining in California, which used giant water cannons to wash hills down and funnel the soil through sluice boxes to remove gold, sending massive amounts of mud "wastes" downstream. Such willingness to demolish a living landscape for its hidden bits of gold was shortsighted, and hydraulic mining was eventually stopped in California because of its impact on cities and farms downstream.

While coal currently is used to generate 45 percent of U.S. electricity, Appalachian MTR coal provides, at most, 4.5 percent. We discussed energy alternatives. Windmill generators have been proposed to make electricity while helping the economy in West Virginia. "Some say windmills kill birds and bats and make a funny noise. Hell, this here," Larry said, pointing to the devastated scene in front of us, "kills everything. You don't like a windmill, you can take it down, but if you don't like a mountaintop removal mine, you can't put the mountain back."

Larry started telling people about mountaintop removal in 1986, figuring back then that once people saw what was happening, it would immediately stop. Twenty years later, after showing his place to more than 14,000 people, his dedication to the mountains in the face of intimidation from Big Coal and many personal threats, is an inspiration. Yet MTR continues, and that is heartbreaking.

"In the most powerful country in the world, so they say—the most educated country in the world, so they say—coal was once essential because we were a young country; like a child," Larry told us. "Now we're walking upright. Now we're powerful and educated. Why are we still using the most barbaric form of energy known to mankind?" He answered his own question: "It only takes two and a half minutes in a power plant to burn a ton of coal that sells for $245. Think how much money they're making. It's not about energy or jobs. It's about making a profit."

The last thing Larry told us on that day on Kayford Mountain was that a senior citizen march protesting MTR coal mining would begin at the state capitol the next morning. Larry was going to join the group and invited us to come along on that first morning of the five-day walk.

The West Virginia state capitol dome gleamed gold in the background as leaders from Climate Ground Zero, Mountain Justice, and the Mountain Keepers Foundation addressed the gathering walkers.

Coal was in the newspaper headlines that morning. West Virginia senator Robert Byrd had surprised the coal industry by taking Massey Energy to task for refusing to help pay to move the Marsh Fork Elementary School, the one Larry had described as threatened by a coal slurry reservoir above the school (Ward, 2009).

Organizer Roland Micklem, 81 years old, described the message he hoped the group would deliver in the coming days: "Coal companies do not own the mountains; God owns the mountains." He planned to seek an audience with Massey authorities at the end of the march. The group expected to face harassment, particularly as they approached the mine.

Some in the group carried signs and banners that read: "Clean Energy Now"; "The Mountains Belong to God, Not King Coal"; "Elders Say No to MTR"; and "How Much did King Coal Pay for Your Soul?" As they walked along, there were occasional honks of support from passing cars. Roland, leaning on a walking stick, told us he felt compelled to take this action. "I'm accountable to myself, and I don't want to look in the mirror, to regard myself, and feel I didn't do everything I possibly could to stop this mountaintop removal."

One of the walkers that morning was Jesse Johnson, who had twice been the Mountain Party nominee for governor of West Virginia. He was not currently running for anything, but explained his motivations for getting into politics and for being there that day: "You used to be able to sit at my farm in Lincoln County and not hear anything except nature. Now, I'm hearing blasting going off. The largest MTR site in the world is coming closer all the time. In the future, I can see myself standing on my great-great-grandparents' grave, where I intend to be buried, and trying to protect it like Larry." His political action career began because every household near his home was hit by high cancer rates, and the residents were using well water contaminated by an industrial dioxin dump.

Jesse addressed the question we had posed to Larry: with so many jobs lost to modernization in coal mining, why do those hundreds of thousands of former workers and their families not support a different approach to the local economy? "The problem is," he said, "they get hornswaggled into believing the myth [that coal mining is good for the region]. The extractive and polluting industries are where all the dollars are, and they purchase more than 75 percent of the advertising in the media, and have come in now and bought the media."

That answer made us recall the billboard message "Coal, the Real Power behind America." Coal culture runs deep and wide in Appalachia. Generations have worked in the mines. One local told us residents either mined coal or flipped hamburgers; there was nothing else. Mining companies have long been paternalistic, providing housing, health care, and a predictable paycheck, but loyalty to coal carries a steep price. West Virginians hope for a future of new opportunities for families to earn a sustainable living. We hope that will happen before all the mountaintops are blown away and shoved into the river channels. West Virginia's state motto is "Wild and Wonderful," but coal's impact on people, land, water, and the atmosphere turns that slogan into a sad irony. Yet much beauty remains that is worth saving, and many West Virginians we met had not given up hope.

In January 2011, the EPA revoked an Army Corp of Engineers permit that had been issued in 2007 for a large MTR mine in West Virginia. The owner of Spruce No. 1 Mine, Arch Coal, had planned to clear 2,278 acres and dump material into six miles of streams near Logan City (37°51'N). The EPA announcement read: "The proposed Spruce No. 1 Mine would use destructive and unsustainable mining practices that jeopardize the health of Appalachian communities and clean water on which they depend. We have responsibility under the law to protect water quality and safeguard the people who rely on clean water" (Malloy, 2011).

Midwestern Rivers and the Population Center of the United States

Coal was not yet done with us as we entered eastern Kentucky. One of the worst coal industry disasters occurred on October 11, 2000, when a dam storing toxic slurry (on an MTR site operated by Massey Energy) gave way and sent 300 million gallons (twenty-five times the volume of the Exxon Valdez oil spill) of black goo downstream through the community of Inez (37°52′N). The sludge moved down Coldwater and Wolf Creeks to the Tug Fork, and some material reached the Ohio River. No human lives were lost, but the EPA called it "the largest environmental catastrophe in the history of the southeastern United States."

In Inez, residents were preparing for a Harvest Festival when we arrived. A young woman selling popcorn had been a high school cheerleader when the flood of coal waste came. "It was thick, black, and gooey," she remembered, "so solid that thrown rocks bounced off of it.

They brought in big vacuum machines to clean it up. Things are clean now, and they widened the creek in case it ever happens again." She finished that thought with a giggle. The idea of it happening again must be unnerving, yet the townsfolk remain.

Through West Virginia and across Kentucky and parts of Indiana and Illinois, we explored a slice of the vast Ohio River watershed: the New and Gauley Rivers, the Kentucky River, and the main course of the Ohio River en route to the Mississippi. Settlers, settlements, transport, pollution, and the rivers' inclinations to flood are all elements in this part of the story.

The look of the forested land changed as we moved westward through Kentucky to Fort Boonesborough State Park (37°56'N). There, beside the Kentucky River, a party of settlers led by Daniel Boone arrived on April 1, 1775. Seeing the advantages the river offered, they established Kentucky's second pioneer settlement. The riverside setting has changed since Boone's day thanks in part to locks built early in the century to facilitate barge travel.

Near Port Royal, Kentucky, we had lunch with author and poet Wendell Berry; his wife, Tanya; and their grown son, Dave, at their farm. The Kentucky River, visible from their house, is part of the special character of the property. "It lies within a hundred steps of my door," Wendell wrote in an essay titled "A Native Hill" (Berry, 2002, 3). Small streams on his property feed the river—which merges with the Ohio River just a few miles downstream—and water figures in many of Wendell's poems and essays. "The creation is musical," he wrote, "and this is part of its music, as birdsong is, or the words of poets. The music of the streams is the music of the shaping of the earth, by which the rocks are pushed and shifted downward toward the level of the sea" (2002, 19).

Wendell is tall and lanky with penetrating blue eyes. Tanya was warmly welcoming and is an excellent cook; she grew up in Mill Valley, California, another 38th-parallel location, and speculated that the similar angle of light might be something agreeably familiar about Port Royal. Wendell became most animated as he drove us around the farm and through his "biscuit woods," containing a mix of black walnut, butternut, ash, and tulip poplar trees that provide good fuel for baking. It was wonderful to spend time with someone whose books and poems so eloquently explore the importance of place and home and sustainable farming.

Lexington bills itself as the "Horse Capital of the World" and, besides its famous horse stables and racetracks, houses the International Museum of the Horse (38°08'N). Although our primary interests along the 38th parallel were water and the environment, an intriguing horse theme had become evident. In China, there were the ancestors of the modern horse, the Przewalski's breed. Turkmenistan's Akhal-Teke horses are ancestors of the Arabian breed. Spain's Andalusian horses originated along our chosen latitude, and the Atlantic Coast of the United States had the island-dwelling Chincoteague ponies. Spanish mustangs run wild just east of our Mono Lake Basin home.

Certainly, different breeds of horses are associated with other latitudes, but a connection between 38°N and grazing animals of all types, including the cattle, yaks, sheep, and goats we encountered in parts of Asia and Europe, is more than coincidence. An article in the *Stockman Grass Farmer* advised ranchers who want to produce grass-fed beef to move close to the 38th parallel: "Almost all of the year-around grass finishing in the world occurs within 100 miles north and south of this degree of latitude. It is far enough North to grow cool-season perennials but far enough South to avoid long periods of deep snow" (Nation, 2010, 24). Latitude also affects temperature; cattle eat less in extreme heat. Holstein females, important as dairy cattle, weigh more when born and when mature above latitude 38°N in the United States. Daily maximum temperatures become limiting in warmer climates, so "at latitudes of less than 34°N in the United States and Caribbean, Holstein females weigh 6 to 10 percent less at birth . . . and average approximately 16 percent lower in weight at maturity . . . than in the northern latitudes" (National Research Council, 1981, 80).

At Louisville, the state's largest city (38°15'N), an interstate highway passes between the city and its Ohio River waterfront, blocking views and access. The freeway remains, but now a lush park, with sculptures and architecture that celebrate water, draws people under the freeway to reach the river's edge to view the bustling river traffic. The city of San

Francisco went through a similar recovery of its bayshore waterfront after an earthquake made it feasible to move a badly located freeway.

Industrialization and urbanization along the Ohio River in each of the eight states drained by its 200,000-square-mile watershed have made it the most toxic river in the nation. As we saw at Chesapeake Bay, storm runoff from diverse sources is today's major challenge. Pollution from point sources such as factories was more successfully managed after the national Clean Water Act was passed in 1972, yet the Ohio River Foundation has become frustrated with lax enforcement in the past decade and has begun independent monitoring investigations to identify polluters and force compliance. That program is similar to the efforts of the NGO Green Camel Bell we saw in China. We were startled to realize that an independent organization was needed to audit and aid government enforcement in the United States, where "the political will to enforce water pollution laws has waned and is now further stressed by economic recession and shrinking government budgets," according to the foundation (Ohio River Foundation, 2012).

Out in the river, opposite Louisville, are the Falls of the Ohio, the only point between Pennsylvania and the Mississippi where boats once encountered dangerous rapids, until the Louisville and Portland Canal Company built locks and canals in 1830. The U.S. Army Corps of Engineers took over the operation in 1874 and enlarged the canals, then added a new dam in 1962. On the Indiana side of the river, one family has run Schimpff's Candy Store in Jeffersonville since the 1850s. Red marks painted on the outside of the storefront mark where the flood of 1937 reached, when the Ohio and Mississippi Rivers left their banks for more than a thousand miles. It was the worst natural disaster in the United States before Hurricane Katrina. Over eight hundred thousand people were displaced from flooded homes, and hundreds died.

A chilly morning brought our first view of the grand Mississippi River when we arrived at Grand Tower, Illinois (37°37'N). Looking for a place to warm up, we found the Mississippi River Museum, being used as an office by County Clerk Charles Burdick, who had navigated 12,350 miles of inland waterways as a riverboat pilot and captain. The Mississippi

River runs from Minnesota to New Orleans, and its watershed extends across 41 percent of the continental United States. The big river is fed through major tributaries, like the Ohio and Missouri Rivers, and smaller ones that reach into thirty-one states, with connections to the Great Lakes and, via the Gulf Intracoastal Waterway, Texas.

Kaskaskia Island was visible on the far side of the Mississippi from Fort Kaskaskia, the only piece of Illinois *west* of the river (37°58′N). The land ended up on the wrong side after the Mississippi River changed course in an 1881 flood. "Kaskaskia, where the West began, Mother of 1,000 Cities, Paris of the West," was the first capital of the Illinois Territory and the capital of the state until 1820, but the flood washed away its physical connection to the rest of Illinois. Our visit to Kaskaskia found only quiet farm fields.

A highway bridge at Chester, Illinois, led us to the beautiful little French-flavored town of St. Genevieve, the oldest town in Missouri. A car ferry also crosses the river there, at *exactly* 38°. Of course, a ride over and back as walk-on passengers was irresistible. The Mississippi was a mile wide at that point and looked exceptionally swift and powerful from a small ferryboat chugging across its surface.

"I'm out planting a forest. Please leave your name and number, and I'll try to get back to you before it matures." That is the answering machine message heard in the opening scene of a video produced by the Pioneer Forest about its founder, Leo Drey.

Wendell Berry had suggested we visit the forest to learn about its half-century-long experiment in sustainable forestry near Salem, Missouri (37°34′N). Forest manager Terry Cunningham and two other foresters told us about the unique program. The vision started with Drey, who began purchasing land in 1954; the total eventually reached 150,000 acres, most from a distillery company that had used the oak wood for barrels. It became the largest private landholding in Missouri and remains open for public use. Drey aimed to demonstrate methods of sustainable logging, without going broke in the process.

The Pioneer Forest follows a long-term cycle of single-tree harvesting that minimizes erosion and runoff on the watershed, makes for an aesthetically pleasing forest with trees of different ages, and improves reproduction and overall health of the trees. The best trees are left in place, instead of the worst, and the forest ecosystem remains relatively undisturbed. This approach contrasts with clear-cutting, favored by the Forest Service in such mixed hardwood forests despite the increased risk of erosion and of regrowth that creates thickets that must be aggressively thinned. An aerial photograph of part of the Pioneer Forest revealed widely spaced stumps within an otherwise untouched canopy of large trees.

Reducing the soil disturbance caused by clear-cutting minimizes watershed impact, and that is one of the Pioneer Forest's chief contributions. The forest staff share their system with other private forest landowners, and also hope to influence policies along the Current River in the Ozark National Riverways Park, which borders the Pioneer Forest. Part of the National Park System, Ozark National Riverways controls only a narrow quarter-mile strip of shoreline along the riverbank. All-terrain vehicles driven into the river and across its banks create erosion. The Pioneer Forest staff would prefer management more in line with standard National Park Service resource protection, which bans destructive activities.

In nearby Edgar Springs, Missouri, we located an intriguing geographic landmark. Each national census identifies the nation's "mean center of population," a rather strange statistic that calculates the point at which an imaginary rigid, flat map would balance if all residents of the nation were of identical weight. When Alaska and Hawaii became states in 1960, the center point moved about two miles farther south and ten miles farther west. For the past several decades, the point has hovered around the 38th parallel, gradually shifting westward and southward. The 2000 Census placed it 2.8 miles east of Edgar Springs, a tiny town in central Missouri (37°41'N; 91°48'W). The town's distinction was noted on a highway sign, but we went into town to see what else might be found.

The lady at the general store said she thought there was a stone marker at the north end of town with information about the designation, but added, "I never go that way." We found the marker outside the gates of a cemetery, where other stones told another local population story.

(The 2010 Census has since documented a continued southwest drift: new "population center" markers will be needed near Plato, Missouri, at 37°30′N.)

More than 80 years ago, during the first years of the Great Depression, a dam was built on Missouri's Osage River by the Union Electric Company to create the Lake of the Ozarks (38°12′N). Once the reservoir was completed in 1931, water extended 180 miles upstream from the dam, with 1,100 miles of coastline, making it the largest man-made lake in the nation. That intricate shoreline reflects the complexity of the Ozark Mountains in Missouri, where the Osage River meanders across the western half of the state. Another major dam was later added upriver, backing water into Harry S. Truman Reservoir. Before all of this dam building, the Osage River had been followed by traders and settlers coming west from St. Louis to Independence, Missouri, to begin traveling the Santa Fe Trail.

Much of that historic route also coincided with the 38th parallel, and with our path westward after the oak woodlands of the Ozarks gave way to prairie on the plains of eastern Kansas.

Seagulls in Kansas and the Santa Fe Trail

The Kansas prairie is, despite first impressions, not perfectly flat, but features low, undulating hills. Prairie grasses, gone dry in early autumn, replaced the green of the Midwest heartland. In eastern Kansas, where 30 to 35 inches of rain fall every year, grasses often grow about five feet tall, though some species send stalks over eight feet high. As we would see, drier conditions in western Kansas bring a shift to shortgrass prairie species better adapted to 10 to 12 inches of annual precipitation. Periodic fires were a key ecological force shaping the evolution of prairie habitat; another force was grazing by Bison (*Bos bison*), the keystone wildlife species in that ecosystem that created conditions for many other species. These large grazing animals gouged wallows for dust or mud baths, which rain would fill with shallow water, providing habitat for frogs and other aquatic species.

Tallgrass Prairie National Preserve was established by Congress in 1996 in the Flint Hills of eastern Kansas (38°26'N). Because of local opposition to constraints that typically come with a national park, the legislation that set up this preserve limited National Park Service acreage to 100 acres; The Nature Conservancy owns the rest of the 11,000 acres.

Bundled up against cold wind and occasional raindrops, we walked away from the ranch buildings that serve as visitor center and park headquarters. Once the trail dropped down a hill so that no structures were visible, the open sky and an expanse of unfamiliar vegetation took charge of our awareness. Tallgrass prairie plants are unfamiliar to most people because farmland conversion and urban growth left only 4 percent of an ecosystem untouched that once covered 140 million acres in North America. The preserve protects a significant remnant of tallgrass prairie, though Bison were missing during our visit.

Just a week later, that lack was addressed when thirteen Bison were brought in from Wind Cave National Park in South Dakota. The following spring, the first Bison calf in 140 years was born on that remnant of tallgrass prairie. Two more births followed. Local cattle ranchers had voiced concerns over disease that might be passed to their herds from the Bison, but other parks such as Wind Cave and Yellowstone have demonstrated how to manage herds to avoid such problems.

Re-introducing Bison was only one of many contentious issues before the preserve's creation, as William Least Heat-Moon explains in his book *PrairyErth*. The book focuses on Chase County, Kansas, where Least Heat-Moon explored the "center" of the United States, located by tracing diagonal lines from the four corners of the lower forty-eight states—a calculation that also coincides with the 38th parallel.

"One would as soon expect to find sea-gulls in Kansas," Mark Twain wrote after visiting Mono Lake (Twain, 1873, 268). We discovered there *are* seagulls in Kansas, lots of them. Also cormorants, pelicans, and phalaropes. Central Kansas has the largest inland wetlands in the country. It was a treat to cross miles of prairie and arrive at Quivira National Wildlife Refuge (38°12'N) on the sunny day the staff celebrated National Refuge Week. Blue skies had returned, and many familiar birds were out on the slightly saline water. Quivira and neighboring Cheyenne Bottoms provide critical habitat for migrating birds on large expanses of surface water in the prairie landscape.

Biologist Rachel Laubhan and Ranger Barry Jones welcomed us to their event, shared much information, and fed us a roasted-pig dinner. At sunset, on the northern edge of Quivira, flocks of White-Fronted Geese (*Anser albifrons*) flew in for the night, ducks were settling, blackbirds massing, and all were honking, quacking, or singing.

The next morning, we stopped in at the new Kansas Wetlands Education Center at Cheyenne Bottoms, a showplace for environmental education and a joint effort between Fort Hays State University and the Kansas Department of Wildlife and Parks. A sign for the Western Hemisphere Shorebird Reserve Network brought back memories of installing the same signs years ago at Mono Lake.

As we saw in China, Turkmenistan, Turkey, Spain, and Portugal, wetlands compete with farms in Kansas for a limited water resource. The local watershed serving Cheyenne Bottoms was called Rattlesnake Creek Basin. On the portion managed by The Nature Conservancy, TNC had fought a legal battle with local farmers to ensure adequate water reached the wetlands. The conservancy based its argument on the doctrine of public trust, which recognizes the government's legal responsibility to protect resources that have environmental and aesthetic value—resources held in trust for the public. To avoid that kind of lawsuit in the future, the U.S. Fish and Wildlife Service negotiated the collaborative "Rattlesnake Creek Partnership," intended to keep refuge lands saturated without putting farmers out of business. Mark Sexson, a public lands supervisor with the Kansas Department of Wildlife and Parks, explained that, instead of cutting a farmer down from 16 inches of water a year to only 5 inches, "which no one can use to raise anything, they're given 25 inches to use over five years. So if they want a high-water-use corn crop, they may use 16 inches of that 25 in one year. Then they may decide to not irrigate for several years. Or they can grow wheat, which doesn't take very much water, and use 5 inches a year to irrigate for five years."

When we visited, it sounded as if the well-intentioned collaboration was floundering, but the objective to stretch limited water supplies between agriculture and the wetlands environment was worth pursuing with such creative approaches.

Heading out of Great Bend, Kansas (38°21'N), we were following the Arkansas River across the plains of central and western Kansas, coinciding with the 38th parallel all the way to eastern Colorado. The Santa Fe Trail closely paralleled the Arkansas River (pronounced "Ar-Kansas" locally)—the most reliable surface water source in that region. From 1822 to 1880, traders and settlers used the trail to connect Missouri and the East to New Mexico, with a mountain branch leading into Colorado.

Pawnee Rock was a major landmark for Santa Fe Trail travelers because it rose above the flat prairie and marked the approximate halfway point to New Mexico. Hundreds of pioneers carved their names in the sandstone. Although much of the rock was quarried in the late 1800s, stairs to an elevated platform gave us a long, unbroken view across the prairie from its former height.

Not far away was Fort Larned, on the Pawnee River fork of the Arkansas River. In 1859 alone, 2,300 people, 1,970 wagons, 840 horses, 4,000 mules, 15,000 oxen, 72 carriages, and more than 1,900 tons of freight left Missouri for New Mexico. Because some livestock and freight were lost to Indian raids, the fort was established and two companies of U.S. Infantry arrived there on October 22, 1859 (*Tiller & Toiler*, 2009).

Today, gigantic cattle feedlots surround the famous Queen of the Cowtowns, Dodge City, where statues of Wyatt Earp and a longhorn steer celebrate the era of cattle drives from Texas, long treks led by thirsty, rowdy cowboys who were themselves corralled by gun-toting lawmen in the railhead town. Just east of Dodge we crossed the 100th Meridian, or 100°W latitude (at 37°48'N). Beyond that point, as we continued westward, we would increasingly appreciate the truth in John Wesley Powell's description of the "line between the humid east and the arid west," marking where less than 20 inches of rain fall each year. In that region, Powell wrote, "agriculturalists will early resort to irrigation" (Powell, 1879, 2–4). Suddenly, our hair, which had curled in eastern humidity, fell straight again, and chapped lips felt the need of lip balm. Across western Kansas, irrigation waters corn and grain that becomes food for cattle that in turn feed much of the nation. The first cattle feedlots, where animals finish fattening before they are slaughtered, were established at Garden City, Kansas, in the early 1950s.

A short drive west of Dodge, Garden City is home to the 3,670-acre Sandsage Bison Range, where state park manager Tom Norman showed us the herd of ninety Bison. Their shaggy bulk reminded us of the yaks of western China. (They are in the same genus, and Canadian ranchers have interbred them to produce "Yakalo.") Bison have been raised here ever since "Buffalo Jones," one of the founders of Garden City, rescued a few of the last remaining animals from extinction. Offspring from fifty-seven calves that he brought to Garden City make up the herd south of town, and more significantly, became the source of new Bison herds around the world.

A few scattered juniper trees on the preserve—planted in hopes they would modify the climate and actually increase rainfall—were evidence of a quixotic attempt to create and sustain a forest on that sand-sage prairie. The 30,000-acre Garden City Forest Reserve was established in 1905 with initial tree plantings of juniper, mulberry, and locust trees. In 1908, another 272,387 acres were added, extending west all the way to the Kansas-Colorado border. With that addition, the reserve became the Kansas National Forest. By 1915, millions of tree seedlings had been planted, but climate and soil realities doomed the project. The trees were not suited to the water and soil conditions that served that natural prairie grassland.

While driving through Bison pastures, Tom pointed out the cooling towers of a coal-fired power plant visible in the distance. Needing lots of water, the power company had purchased 20,000 acres with water rights formerly used for crop irrigation. Controversy ensued, but Tom said: "To me, and most of us in the wildlife field out here, we thought that was the most positive thing to happen in southwest Kansas for wildlife in 50 years. I was pretty excited about seeing so many [irrigated crop] circles going back to permanent cover. One of these days, it's not going to be economical to pump water and put it on crops anymore. What's going to happen when it isn't? Are they just going to turn off the wells and back away?" Tom was referring to the groundwater pumping issue that dominates the environmental and economic future of that region: overdrafting of the Ogallala Aquifer.

Mining the Ogallala Aquifer

The Ogallala Aquifer extends from Nebraska down to Texas. This enormous source of deep groundwater allows the region to grow corn and grain and raise most of the nation's cattle on high plains that became highly productive only after efficient groundwater pumps and center-pivot irrigation sprinklers were invented.

At the Kansas Groundwater Management District No. 3 in Garden City, Executive Director Mark Rude, dressed casually in blue jeans and a plaid shirt, somehow stays upbeat as he faces undeniable facts: there is less than an inch of recharge into the aquifer each year, but in places withdrawal is happening hundreds of times faster than that. Many pumps now work at reduced capacities. When the water runs out, the economy and environment of this vast region must change.

Meanwhile, in the attempt to slow, though not yet to control, the situation, each water user in District 3 is limited to a set allotment enforced with well-monitoring meters and sensors. The gloomy news was that the groundwater depletion had not been stabilized, because the limits on pumping were inadequate. "Economists are winning, not hydrologists," Mark told us.

By drilling test wells, early ranchers became certain there was a vast, "inexhaustible" underflow, but most of it was too far down for early pumps to reach. Garden City got its name because its windmills tapped into a fairly shallow water table to irrigate gardens and fill surface ponds. "When deep turbine pumps came in," Mark said, "that opened up the range land; you could lift from any depth." Local wells reach down 60 to 70 feet, but most wells tapping the Ogallala Aquifer descend several hundred feet, and farther south, along the Oklahoma border, some have chased the declining water table down 700 feet. The cost of energy to lift water from those depths becomes a limiting factor.

Center-pivot sprinklers were the other key innovation. An aerial photograph of the groundwater district, mounted on the wall of the conference room, showed mile after mile of green circles fed by ten thousand wells. "The center-pivots allowed massive development in the 1960s and into the '70s," Mark said. "Wells could be put right beside the sprinkler pivots. You could take $50-an-acre ranch ground and make it $1,000-an-acre ground." A state engineer during that era approved pump permits without regard for sustainable yield. Saying "no" to one rancher's new wells when neighboring ranchers were pumping was politically difficult. The easy choice led to too many pumps tapping the aquifer.

We asked Mark to predict the future of western Kansas, based on the trends he saw. "I'm an eternal optimist," he said. "I think conservation of the water supply will increase. There will be a lot more managed recharge, a lot more accounting for every drop." But then he described the problem between hydrologists and economists, adding, "In eastern Kansas, with more water, they can manage for sustainability, but in western Kansas, our long-term plan is to deplete the aquifer."

Mark knew about the decision, based on the public trust doctrine, to protect Mono Lake by reducing stream diversions. "But that likely wouldn't have happened if the water was below ground instead of on

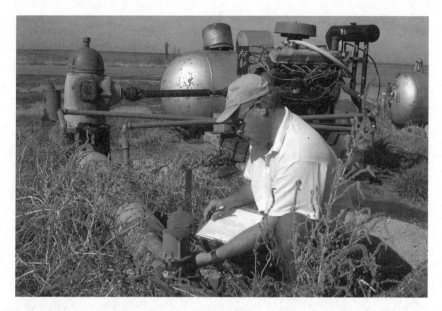

Figure 22. Jason Norquist checks groundwater well monitors to ensure that farms comply with limits on pumping from the overdrafted Ogallala Aquifer in southwest Kansas.

the surface," he said. Effects on Mono Lake, including the loss of half its volume and a drop in its surface of 45 feet, had become strikingly evident. Tufa towers that had been submerged were exposed; a land bridge formed out to an island, allowing coyotes to rout out a gull rookery; and dust storms began blowing off the exposed lakebed. Effects on an underground aquifer are out of sight, and so they are out of mind.

"We have 10,000 flow meters out there on wells. Sometimes those meters seem to break down at the most inopportune time from a compliance standpoint," Mark said, wryly. "Our answer to that problem is a temperature logger, an inexpensive device to strap onto a pipe that records when the well is operated, because groundwater is cold."

To show how the monitoring works, Jason Norquist, an assistant manager, took us out to watch as he installed temperature sensors on several wells. As he drove, he explained how he knew when one farmer had been playing games with his meters. "Meters were running backwards because he had them installed backwards. The patterns where there are leaks, of algae and moss on the pipes, didn't match up." Washers and nuts had lost their rust when a wrench was used on them. Jason found five more wells at the same ranch with meters on backwards, so he called the farmer up and told him what he had found. They went over that farmer's pumping hours, and since then, the farmer's attitude had turned around.

"Most of these guys keep track of new innovations," Jason added, "and are bringing in drip irrigation systems and more efficient ways to use the water. One of the first in the area to put in subsurface drip took his water use from 25 inches per acre down to about 14 or 15 inches. Then his dad and his grandpa . . . well, they grew up the old way. They like to *see* the water. With that subsurface drip they can't see any, so whenever he shut his well off, they'd come out and turn it back on. There's a generation gap."

As Jason installed a temperature sensor on a pump in a field of corn stubble, he explained that the old practice of flood irrigation extended across entire 160-acre quarter-sections right to the corners. With center-pivot sprinklers, the resulting circles reach only about 125 acres. Drip irrigation is more efficient per acre, but also allows farmers to plant out to those corners again, so overall, water savings from the more efficient system are not as high per quarter-section as might have been hoped.

That sounded like California's Central Valley Project, whose goal to replace groundwater overdrafting with imported water was not achieved, partly because the new water was instead used to extend irrigation onto new acreage.

At breakfast, we spoke with Mark Sexson, the regional public land supervisor with the Kansas Department of Wildlife and Parks. His responsibilities cover just about anything to do with water in western Kansas. Mark confirmed many details we had been hearing about Ogallala Aquifer overdrafting. He pointed out that "use it or lose it" water rights laws perpetuate a lack of incentive to conserve. When we asked what he saw coming in the future for that region, Mark said: "As water declines in the aquifer, rather than just shut off water to people who had water rights—that's harsh—instead we should pay to retire the rights. Then put the land back into grass to perpetuate prairie chickens. As far as wildlife goes, the best thing that can happen is to run out of water fairly quickly," so that the wildlife habitat in the sandhill prairie can be restored.

Recharge for the aquifer in Kansas originates uphill in Colorado. The State of Kansas took Colorado to court over holding back too much water, and won. Arkansas River water stored in an eastern Colorado reservoir now must be made available to Kansas. Mark feels that a part of society's problem is the "disconnect, whether it is wildlife or farm or water issues, between the urban vote, the people who have the power in legislatures passing laws, and the reality of life in the rural areas, where food comes from." Environmental education was important, we suggested.

Mark agreed. "My wife, Kathy, directs the Center for Conservation in Garden City. She gets one hundred thousand visitors a year, but even in Garden City they don't understand where their water comes from."

As Mark suggested, urban demand was another part of this Kansas water story. We met with Garden City mayor Nancy Harness in a downtown coffee shop. An impressive leader of a cosmopolitan community, she began by talking about those problems of perception and long-range planning.

"There's a mindset that water is here to be used and acceptance that someday it will be gone," Nancy said. "In this part of the country, things seem so temporal. You can drive around here and see these little communities with remnants of homesteads, houses, and churches. Our culture is 'use it and move on.'" When that cultural pattern dominates, it becomes nearly impossible to effectively address long-term water-supply issues. Our global travels had shown us, time and again, how people's strong connections to a place provide the motivation to create sustainable lifestyles and mount successful resistance to short-term extractive policies driven by impersonal corporations or distant politicians.

As for water awareness and conservation, Nancy said, "we're too green." She did not mean "green" as in "environmentally conscious," but rather the color itself. "Everybody loves to water their lawns, and they water them, the sidewalk, and the street. Which is maddening because it is a finite resource. We [the city] did adjust the pricing so the more you use, the more you pay, trying to make average water bills reasonable. But we have yet to get truly serious about water. No one has ever spoken to . . ."

"Limits?" we suggested.

"Yeah. We're just in complete denial. There is no doubt about it. [People feel] you can turn it on, it's always there." She told us about a neighbor whose sprinklers ran every day, rain or shine. We mentioned that smart sprinkler systems are available now that monitor moisture and decide whether to come on or not. Nancy laughed. "I don't think it's the sprinkler's fault."

The biggest employer in the Garden City region is the Tyson Foods Packing Plant. Earlier that day, Nancy had been on a tour of the cattle-processing facility, where 3,100 people work. "They kill four hundred cows an hour—thousands of cattle every day—and contribute millions of dollars every month to our economy. It's huge. The cattle all come from within 100 miles of the plant. All the corn you see around here fattens the cattle at the feedlots, which then go to the packing plant. Hundreds of gallons of water are used to process each carcass." To secure the water they need, Tyson purchased thousands of acres including water rights (as the local coal power plant had done). Those acres also gave them a place to send their reclaimed wastewater, which fertilizes crops.

The plant attracts economic refugees to Garden City because, until recently, they could always find low-skilled work. "There are twenty-seven languages spoken by people at the plant," Nancy said. "It is the 'United Nations.'" Folks can come here, earn $12 an hour, and not have to speak English. But they're awful jobs. They come from Somalia, Burma, . . . and Michigan," she finished, laughing. "We had twenty-five teachers who came here from Michigan this year. I spoke to some of them and told them 'I'm sorry things are not good in Michigan and I'm glad you're here. I promise you, in February, *you'll* be glad you're here, when it's cold and snowy up north.'" Nancy welcomes this mix of cultures as a positive thing for her community, though it seems likely that the influx of newcomers is undermining long-term commitment among some residents in that locality.

Parallels between the dilemmas of unsustainable pumping from the Ogallala Aquifer and mountaintop removal coal mining are striking: in each case, a short-term economic choice will leave future generations with enormous problems and environmental loss. It was another hard story to hear and another motivation for spreading awareness about where the water sources are that grow our food and send the life-giving elixir to our taps.

The Ogallala groundwater basin extends into neighboring Colorado. We noted the changing appearance of the Arkansas River as we drove: wet with emergent groundwater in Great Bend, bone-dry in Garden City, but perking up as we followed the mountain route of the Santa Fe Trail across the state line.

Colorado, the Headwaters State

The Arkansas River's headwaters are within Colorado, named the "Headwaters State" because so many rivers originate from the state's mountains. Water from Colorado reaches into eighteen states and Mexico. Some snow that melts west of the Continental Divide, after passing through dams and aqueducts, emerges from faucets in Los Angeles.

The southern Rocky Mountains divide into a complex series of ranges with 14,000-foot peaks that made it impossible to stick closely to the 38th parallel as we headed west. In the following days we swooped north and south and north again through the Sangre de Cristo and San Juan Mountains, seeing some of Colorado's most scenic canyons, spectacular passes, and historic mining towns.

West of the city of Pueblo, the Arkansas River canyon climbs upward and the river becomes a swift mountain stream. We reached Salida,

Colorado, 7,000 feet above sea level, on a night of early winter weather and, hosted by former Mono Lake neighbors, took a day off from traveling. There we learned about a battle being fought by local citizens against Nestlé Waters' plan to export local water to a Denver bottling plant. The concerned citizens we met that day were still reeling from a decision reached by their Chafee County commissioners a few weeks earlier. Land-use and pipeline permits had been approved, allowing Nestlé to move 65 million gallons of water a year to an Arrowhead brand bottling plant. Twenty-five trucks would make the 240-mile round-trip to Denver each day.

We met Vicki Klein and Betsy Steinberg in a cozy coffee shop near the river as rain and snow kept falling, and later joined John and Ann Graham, who are officers of Chafee County Citizens for Sustainability, the group that led a campaign against the bottled-water proposal. Their opposition was based partly on the number of trucks added to local highway traffic, and partly the illogical waste generated by bottled water. "Taking the water doesn't provide local jobs," Betsy added. "They aren't even gassing their trucks up here." The county will receive about $80,000 a year, "but that's not much money considering the problems this is going to generate. They treated us like we're a bunch of stupid locals, asking, 'What if we put on the bottle label that it is from Chafee County? Would you like that?' It was insulting."

The carbon footprint of bottled water is amazingly high. In the United States, 28 million bottles of water are sold each year in plastic containers that require 17 million barrels of oil. The manufacturing process consumes three times more water than what makes it inside those bottles. Text on a metal bottle made for the Citizens for Sustainability reads: "I bottle my own water."

John's hope was that "the public will just open their eyes and close their pocketbooks [to] companies like Nestlé. Bottled water is passé; bottled water is foolish; bottled water is 'not cool.' It is a last resort for emergencies only."

We were ready for a more positive story, and found one in the San Luis Valley after driving across a snowy pass that brought us back to the 38th parallel near the dramatic Great Sand Dunes National Park.

The site was upgraded from a national monument to a national park after The Nature Conservancy (TNC) purchased local ranches, foiling a water export scheme meant to serve development along Colorado's Front Range, while also creating a new national wildlife refuge. The "Front Range" is what Coloradans call the land east of the Rockies where Denver and its thirsty suburbs are located.

The background to the San Luis Valley water story was given to us by Paul Robertson, a white-bearded local project coordinator with The Nature Conservancy. We met at the Scenic Country Store Cafe, where the cinnamon rolls were a day old, but the waitress promised they would be good when warmed up with butter, adding, "If you don't like it, I'll feed it to the chickens." Coffee money, whatever amount one felt like paying, went into a donation can.

The ancient San Luis Valley is the largest alpine valley in the United States, with over 5,000 square miles of valley floor; the State of Connecticut could fit comfortably within it. The valley was once covered by a hundred feet of water until, during a particularly wet era, water broke out to the south and formed the Rio Grande. Beneath sediments from the San Juan and Sangre de Cristo Mountains lie two aquifers— a deep confined one that generated "head pressure" to push artesian springs to the surface, and above that, separated by a clay layer, another shallow one.

The first diversion ditch tapped the Rio Grande to irrigate farms in 1835. In time, enough water was taken to dry up the river. Ranchers started drilling wells in the 1930s, a process that continued until just recently. Wells now serve 2,500 center-pivot sprinklers for irrigating crops. Most tap the shallow, unconfined aquifer, but enough extend down to the lower aquifer to reduce the pressure and stop the natural artesian flows that people relied on in the early years. Today, forty thousand people live in the valley.

In the early 1970s, American Water Development, Inc. (AWDI), owned the Baca Ranch on the east side of the valley. As their name took no pains to hide, "they were really most interested in the water to move it to the Front Range," Paul said. "At the time, the state didn't control groundwater; almost all Colorado water law was based on surface water." (In California, groundwater is *still* not regulated by the state.)

AWDI sold to American Stockman, backed by investment money from outside the region. "Money kept the battle going," Paul said. "One very flamboyant local guy was leading the effort to export water. We identified this valley as having high conservation values, both biological and natural resource values: the sand dunes, the water, the aquifers. Most ranchers and the land management agencies were very much behind us, because they were seeing water tables drop and springs go dry."

In 1999 TNC bought the 100,000-acre Mendono-Zapata Ranch, the biggest purchase deal they had ever made. The Clinton administration supported the plan to upgrade the site to a national park and also create a 60,000-acre national wildlife refuge. TNC made the purchase and then sold the land to the federal government, so that the former national monument, which had about 40,000 acres, now met the acreage criteria for designation as the Great Sand Dunes National Park.

"Everyone got what they wanted, except the people who wanted to sell water to the Front Range," Paul said. "That one stockholder made an interesting argument. He asked why it was OK to put water into potatoes and ship them out of the valley, but not OK for him to pump the water and ship it directly. Of course, if that had happened," Paul added, "he would have been a millionaire, and no one else would have gotten a nickel out of it. The other way the money is scattered around the valley."

Since 2000, Colorado has strengthened its laws controlling groundwater. No new permits are issued now except for domestic use. There are thirty monitoring wells in the national park and hundreds across the valley; if the water table drops too low, pumping must stop. "That happened in the 2002 drought," Paul said. "To maintain the aquifer, they closed down one thousand wells. Two years ago, they required meters for everybody. Now everybody has to pay attention to groundwater augmentation. Long-term sustainability is the goal." To make it clear that the new policies and regulations would be enforced, a few ranchers who refused to install meters were levied heavy fines.

Colorado's realistic approach to long-term sustainability contrasts with the gloomy forecast for the Ogallala Aquifer. The difference in effectiveness may stem from the San Luis Valley's greater compactness and confinement as compared to the Ogallala, which spans so many states and groundwater jurisdictions.

Paul was also working on conservation easements along the Rio Grande, which flows across the southern edge of the valley through the Alamosa National Wildlife Refuge. At the nearby Monte Vista National Wildlife Refuge, we saw Sandhill Cranes (*Grus canadensis*) that had just arrived on their annual fall migration. After landing, they pointed their beaks to the sky and pranced in place. With a success story to share from the San Luis Valley, we felt like celebrating too.

At Wolf Creek Pass, after the highway climbed out of San Luis Valley, we reached the Continental Divide and entered the region where waters drain toward the Colorado River and the Pacific Ocean, instead of the Atlantic. If the gravel road to Summitville Mine had not been snowed in, we might have turned off, but the mine was 16 miles south of the paved highway at 11,644 feet above sea level (37°26′N). Since December 1992, when the Canadian mining company Galactic Resources filed for bankruptcy and walked away from Summitville Mine, it has been an Environmental Protection Agency Superfund cleanup site. Like most modern gold mines in the West, Galactic processed crushed ore in a "heap-leach" operation, piling it on top of a vinyl liner to be washed with cyanide to extract gold. At Summitville, the liner under a 160-million-gallon pond full of cyanide solution developed leaks. A dozen spills, totaling more than 86,000 gallons, entered Cropsy Creek, a tributary of the Alamosa River, which eventually enters the Rio Grande. Fish died. Everything died. Faced with a $20 million cleanup job (the costs have since grown), and lenient national mining laws, Galactic left it to the federal government to clean up the cyanide pond and a major acid drainage problem at taxpayer expense.

This sad event actually contributed to the preservation of a historic mining town called Bodie, about 15 miles north of our home at Mono Lake. The same corporation, Galactic, had been preparing to mine the hill above the Bodie State Historic Park, where structures date back to the 1870s. A campaign led by the California State Park Rangers Association was under way in 1992 to save the historic ghost town from the impact of a modern industrial mining operation. When the mining company declared bankruptcy, the state park system was able to purchase the hillside, and those who cared about Bodie breathed a sigh of relief.

Our winding route passed through Durango, then swung north up a scenic valley full of aspen trees and back into the San Juan Mountains, through Silverton and Ouray, around another mountain range, and up a box canyon to Telluride, a ski town on the 38th parallel that was a mining center in the late nineteenth century.

Victorian-style houses and a historic downtown district contrast with modern condominiums, while San Miguel Creek cascades along the edge of town. We planned to look into water issues left over from the mining era, then learned that a "350 Global Climate Change" event was scheduled for the day we arrived. That day, in more than 5,200 locales in 181 countries (including every single nation and state touched by the 38th parallel), people gathered to highlight the goal to reduce carbon dioxide levels in the atmosphere from the current elevated level of 392 parts per million back down to 350 ppm, a level considered a safe upper limit to avoid runaway climate change. Telluride shares concerns about the effects of global warming on snowpacks with the ski industry around the world. Organized by the New Community Coalition, bicycles, solar vehicles, and hybrid cars were gathering in the ski town to form the numerals 350.

The organizers needed another car to complete the figure, so our hybrid ended up front and center in the picture, forming the top of the number 5, in front of the courthouse podium. It was nice to briefly become part of that community and share a small role in that worldwide day of awareness, meant to address the ramifications of climate change we had learned about everywhere we traveled.

Colorado River Cleanup and Groundwater for Las Vegas

Entering Utah, we came downhill to the Colorado River at Moab, Utah (38°34′N), famous for mountain biking on slickrock trails and as a place to start river trips. The Colorado River watershed covers one-twelfth of the area in the Lower Forty-Eight, running through parts of Wyoming, Colorado, Utah, New Mexico, Arizona, and California, on a 1,440-mile trip to the Gulf of California in Mexico. Like the Yellow River, heavy loads of silt gave the river its name, the Spanish *colorado*, for the reddish color of its water.

Where reservoirs restrain the modern Colorado, the silt becomes trapped behind dams, and the river loses its typical coloring downstream. A few years earlier we had canoed for a week on the Green River, through Canyonlands National Park to the confluence of the Green and Colorado Rivers—yet another landmark found by the 38th parallel. Just downstream, Lake Powell, the reservoir behind Glen Canyon Dam, begins. Below the Grand Canyon, Lake Mead, behind Hoover Dam, impounds water for 18 million southern Californians, is the key reservoir serving southern Arizona, and provides almost all of the water supply of the city of Las Vegas in Nevada. We were in Moab to learn about the cleanup project under way to haul uranium mill tailings out of that critically important Colorado River watershed.

Between 1956 and 1984, dozens of uranium mines operated 10 to 20 miles south of Moab in outcrops of the Morrison and Chinle sedimentary layers (clustered very close to the 38th parallel). They sent their ore to a processing facility at Moab because the Colorado River was the only water source in the region that could provide the consistent supply needed for milling. For 28 years, a tailing pile with radioactive uranium waste and high ammonia concentrations grew to cover 130 acres on the site; the edge of the pile was 700 feet from the west bank of the river, except during high runoff episodes, when the river sometimes touched that toe.

"The pile actually started as a slurry pond and over time filled up with more and more sediment. It's kind of like a jelly donut," Elizabeth (Liz) Glowiak told us, "where all the fine materials traveled in toward the middle of the pond, with the sand on the outside. We call it the 'slime' in the middle." Liz and her co-worker, Ken Pill, are geohydrologists. Wearing bright-yellow reflective vests, they took us on a tour of the Superfund cleanup site. "The tailings are wet," Ken said, completing Liz's thought, "because they used up to 1,600 gallons per minute of water diverted off the Colorado River to process the ore." Ken and Liz, who worked for a technical assistance team contracted to the Department of Energy, smoothly took turns providing facts as we toured the Moab Uranium Mill Tailings Remedial Action Project.

From a viewpoint above the tailings, we watched dozens of trucks and loaders at work. Contaminated ground was being loaded onto trucks, which were thoroughly washed before transferring their loads to uncontaminated trucks (which never enter the pit); those loads then were delivered to train cars. Special mechanics were assigned to work only on the dirty trucks inside. Twice a day, a loaded train hauled the

Figure 23. Uranium mill tailings are hauled away from the west bank of the Colorado River at Moab, Utah.

contaminated material 30 miles north, for reburial outside the Colorado River watershed.

"Was the tailings pile lined when they were operating?" we asked.

"No, and that was typical practice in those days," Ken answered. "To give the Atlas Mineral Company a little bit of credit, a lot of other mining sites and milling sites were discharging directly into the river."

With millions of people downstream in Arizona, southern Nevada, and Southern California relying on Colorado River water, if floodwaters washed over the tailings, or if the plume moved underground into the river, the consequences would be devastating. Discussions about how to solve the problem had gone on ever since the mill company went bankrupt to avoid paying for the cleanup (the same tactic used by the mining company at Summitville, Colorado). "We took control of the site in 2001," Ken said. There was only enough money to consider capping the pile in place or to perhaps do some remediation. In 2010, once federal economic stimulus funds became available, the aggressive cleanup project began in earnest to provide a permanent solution. It will take ten years of constant work to move 16 million tons, a literal mountain of tailings.

The mill used anhydrous ammonia to process the uranium ore—"tons of ammonia," Ken said. It became the main contaminant in the groundwater, and a concern for fish habitat along the Colorado because fish take up ammonia in their gills.

It was counterintuitive to learn that low water flows in the river actually increase the likelihood of contaminated groundwater moving toward the river. High-water episodes seep into the pile and create a countertension against the groundwater while diluting contaminates.

Ken and Liz monitor test wells across the edge of the site closest to the river. Besides uranium and ammonia, they sample for copper, selenium, manganese, and total dissolved solids in the brine, pumped from different well depths, and analyze their samples with equipment at the site. Wells also provide a way to proactively skim contaminates when concentrations are high. An "air stripper"—a high-pressure water stream shooting straight up in the air—converted ammonia to nitrogen gas. That required high temperatures and low humidity, which was exactly what the local desert climate provided.

At the far edge of the property, we walked through a riparian habitat that gave refuge to native Razorback Suckers (*Xyrauchen texanus*) and Tui Chub (*Gila bicolor*) fish. Birds were singing within a strip of willows that could become part of a restored natural site at the gateway to Moab, after the tailings are gone.

"Will you be here for the entire ten years?" we asked Ken.

"I'd like to think so," he said. "I was down here when they first took over the site in 2002 and when they put in the first set of extraction wells. Now we're at the point where we're starting to do more remediation. So I'd like to follow through on the whole thing."

In the middle of Moab, at the tents-only "Up the Creek" campground, we met the campground owner, Kimberly Schappert, who came to Moab over twenty years ago. She watched Moab change from a small mining town with a Uranium Cafe and Atomic Grill into a mecca for outdoor recreation, and became involved in county government, working for years to get the huge uranium tailing pile cleaned up. "That entailed meeting with the DOE [Department of Energy] people and the state DOE as stakeholders," she said, "and we would meet with the Metropolitan Water District of Southern California. We flew people out and took them on tours. Eighteen million Southern Californians and a bunch of Arizonians are what gave us the clout."

Money became the biggest stumbling block. "The original company, Atlas, had only a $6 million reclamation bond," she said. "Yet the cleanup project is up to $350 million now, and will be half a billion dollars by the time it is done. Atlas went bankrupt, so we taxpayers are footing the cleanup bill."

Kim had lots of ideas for what could be done on the land once the cleanup is finished. "This has been so ugly, but it should become the great gateway to the community." She envisions a multiuse area, with "green" office space, parking for Arches National Park, and trailheads for mountain biking and walking. "This is a recreation area for people from all over the world, so whenever I apply for a grant, I keep in mind that we're not just serving the people of this county, but 2.5 million visitors a year from all over. This is their impression of the United States, so I feel like it's tax money well spent."

"What do you think it takes to see a problem like that and fix it? Is it key people in the community?" Janet asked.

"I think it is. People that have an issue, a cause, and are willing to fight for it. All the great things going on around here are not because of the government. It takes people . . . "

"People like you," Janet said.

"I have to admit, I work my butt off for this community, but other people have done that also, for other causes and issues. That's what gets it done. I'm sure that's what gets it done everywhere."

"You've inspired me to go home and try to make some things happen," Janet said.

Water and gravity have sculpted colorful sedimentary rock formations across southern Utah in some of the nation's most spectacular national parks. Our route included Arches National Park, Canyonlands National

Park, Lake Powell National Recreation Area (behind Glen Canyon Dam), Grand Staircase–Escalante National Monument, Capitol Reef National Park, and Bryce Canyon National Park.

On a bitterly cold afternoon, we arrived in the old mining town of Pioche, Nevada (37°56'N). In 1872, Pioche (pronounced "pee-oach") was the major town of southeastern Nevada. For two years, its population hovered around ten thousand. The boom-and-bust pattern of western mining played out quickly here, as the principal silver mines closed in 1876. Because of frequent gunfights, the town claims to have once been even rougher than Bodie, the notorious mining town north of Mono Lake. Today it is the county seat of Lincoln County.

At the Pioche Library, we met Glennon Zelch, who was heavily involved in an ongoing battle over Las Vegas's plans to export ground-water from eastern Nevada valleys. This current water grab has many parallels to the Los Angeles Aqueduct history that affected Mono Lake and the Owens Valley in eastern California: family ranches being bought up for their water rights, local people overwhelmed by big-city movers and shakers, and water moved hundreds of miles to serve development and growth in a distant, sprawling metropolis.

Glennon is a retired civil engineer who worked formerly with water utilities. When he moved to Pioche from Louisville, Kentucky, he was not aware of Las Vegas's pipeline plans. "I learned it was in the works when I stopped at the building department to see about a permit for my house," Glennon said, "and the lady in there asked me what I did, and I said I was a retired civil engineer. And she said, 'Oh goody, you're the only one in the whole county.'"

Glennon was asked to represent Lincoln County on a cooperative development committee that included a man from neighboring White Pine County and thirty-six people from Clark County, which includes Las Vegas. "So we were outnumbered eighteen to one to start with," he said. "The whole purpose was to figure out a palatable way for the Southern Nevada Water Authority to put in this pipeline without stirring up too much public opposition. I went because I figured I could learn what was going on."

Las Vegas water planners have said they must have a minimum of

100,000 acre-feet of water out of White Pine County to make the pipeline project viable. One acre-foot (about enough water to cover a football field one foot deep) is a sufficient annual supply for one household—or two if used conservatively. "They've bought about 50,000 of that," Glennon said, "and want to file on public land in Spring Valley and Snake Valley to get the remainder." Snake Valley could be the big stumbling block for the project. "If they pump from wells on the Nevada side, they'll pull water out of Utah. If Utah says 'no,' there's going to be hell to pay." If the minimum amount of water required for Las Vegas did not materialize in Snake Valley, the pipeline project would not be economical. Although a tentative agreement had been reached between the states to share groundwater equally in the valley, opponents in Utah remained concerned that depleting the groundwater table would kill vegetation and lead to unhealthy dust storms blowing all the way to Salt Lake City. After the Los Angeles Aqueduct was built, severe particulate air pollution developed in the Owens Valley and around Mono Lake, violating federal air quality standards and leading to expensive remediation for Los Angeles to undertake.

"What do you see as the future for this town and this part of the county?" we asked Glennon.

"How much water you have dictates how big you can grow," he answered. "Los Angeles could not have gotten anywhere near as big as it is if they hadn't stolen the water from elsewhere. Las Vegas is on the same track."

We had heard that Las Vegas was promising a viable economy for the northern valleys, that there would not be "another Owens Valley," because the city would share sufficient water. "Well, we don't have a viable economy now," Glennon responded, "and if we lose our water we can never develop one." The economy of Pioche is today based, not on mining, but on ranching and government services, including the county courthouse, schools, road departments, and a prison. "Without water, we're going to be welfare people. I told the people in White Pine County that I've met, 'File on every drop of water that you've got in your county, because if you don't, they'll get it sooner or later.' Eventually they're going to extend the pipeline even farther north." Something similar had happened at Mono Lake when Los Angeles decided to reach north of the Owens Valley. "That's the logical course of events," Glennon said. "They're not going to say, 'We've taken all the water we're going to take.' They'll just keep wanting more."

He finished with a provocative thought: "To be realistic, you should develop a community up to the limit of its resources. Then go develop a new community somewhere else. You can't just keep glomming all these people into one spot and trying to siphon water in from somewhere else to support them."

Serving the Las Vegas appetite for never-ending growth is another unfortunate example of the focus on short-term resource extraction and economic gain ahead of long-term sustainability, as we had seen in many places. The same philosophy that finds groundwater pumping in eastern Nevada acceptable also justifies the "mining" of the Ogallala Aquifer in Kansas, the removal of mountaintops for coal and destruction of streams in West Virginia, the remodeling of rivers in Korea, and dam building in many nations.

In March 2012, the Nevada state engineer ruled that the Southern Nevada Water Authority can pump 83,988 acre-feet of groundwater a year from the four rural counties in eastern Nevada. That is less than Las Vegas asked for and less than the 100,000 acre-feet that Glennon saw as the minimum to make the project economically viable, so it will be interesting to follow what happens next. An economic recession and major housing slump in Las Vegas was dampening that city's thirst when we visited the state, but more chapters to this story will follow in the years ahead.

Visiting Pioche was a particularly meaningful way to return home, after the around-the-world trip that had led us to so many water issues. The new aqueduct proposed by Las Vegas, that strangest of cities in the desert, mirrored the eastern Sierra history we knew, in which, early in the twentieth century, Los Angeles reached north for Owens Valley water to allow that city to grow beyond the limits of its local water supply. The story brought us full circle, and now, after circling the globe, we were only a few hours away from Mono Lake.

As we crossed the state of Nevada, wild horses grazed near the highway. We watched for aliens along the "Extraterrestrial Highway," but eagerly pushed on. Wind was making waves on Mono Lake when we reached home, piling foam at the edge of the salty, alkaline water and whirling white flecks off into the sagebrush.

PART FOUR ## California

A WATER LINE TO THE PACIFIC

The Giacomini Marsh restoration project alone adds
10 percent to central coast wetlands. That gives you an
idea how little is left.

Superintendent Don Neubacher, Point Reyes National Seashore

Map 4.   California
1. Hetch Hetchy, Tuolumne River; 2. New Melones Reservoir, Stanislaus River;
3. Stockton; 4. Sacramento–San Joaquin Delta; 5. San Pablo Bay; 6. San Francisco
Bay; 7. Point Reyes

Mono Lake to the Sierra Crest

We were camped at 10,430 feet above sea level, just east of the Sierra crest, looking at small glaciers clinging to the north slopes of Mt. Conness and North Peak. Wispy clouds turned pink, and there was the beginning of alpenglow overhead. It was our first evening on a 17-day trek across California, continuing along the 38th parallel from Mono Lake to the Pacific Ocean, exploring the mountains-to-the-sea watershed that sends water to San Francisco Bay and, finally, out to sea.

We approached the California segment differently than elsewhere in the world. At the slow pace dictated by travel on foot, bicycle, and boat, we traced a 38th-parallel "water line" along which battles have been fought over dams, aqueducts, and wetlands, and critical water issues still are being played out. The California list includes Mono Lake, the snow-pack at the Sierra Nevada crest, Hetch Hetchy, New Melones Reservoir, the Sacramento–San Joaquin Delta, San Francisco Bay, Point Reyes, and the Farallones Islands. In each location, the water issues of our home state of California had parallels and contrasts with discoveries we had made at lakes, rivers, marshes, estuaries, and islands along the 38th parallel around the world.

Our starting point at Mono Lake, east of the mountain range, seems hydrologically separate from California's west-slope watershed, until you consider the intricate water system that spans the state. Most Californians are tied together by man-made plumbing, and this ensures that stresses on the water-supply system in one place are today felt throughout the state.

Those connections began in 1905 when the City of Los Angeles asked its citizens to approve a bond for construction of the Los Angeles Aqueduct from the Owens Valley, hundreds of miles to the north. Tapping the Owens River for water allowed the city to grow far beyond the limits of its own local water sources. Thirty years later, the city reached even farther north, adding streams from the Mono Lake Basin to its distant water supply.

The population of Southern California mushroomed even more after the Colorado River Aqueduct began delivering greater quantities of water to the region in 1941, and again after 1960, when voters approved construction bonds for the dams and aqueducts of the California State Water Project. The California Aqueduct carries runoff from dams in the northern Sierra to Southern California cities as far as San Diego, 444 miles away. To the north, the city of San Francisco received congressional approval in 1913 to build a dam on the Tuolumne River in Hetch Hetchy Valley, inside Yosemite National Park. For the East Bay, an aqueduct spanned the middle of the state to tap mountain runoff from the Mokelumne River. During the Great Depression, the federal government began building the Central Valley Project, ultimately totaling twenty dams on five of California's major rivers, plus aqueducts to transport Northern California water to farms in the southern San Joaquin Valley. Today, with more than 1,400 dams and a spiderweb of canals, water choices made by any section of the state may have repercussions hundreds of miles away. Key elements of California's twenty-first-century water web lie along the 38th parallel, including Mono Lake.

At a sendoff party at the lake, we floated in the salty water and then told a group of friends about our hopes for the trek across the state. We had worked as rangers at the Mono Lake Tufa State Reserve for more than two decades, in partnership with the U.S. Forest Service. It took the Mono Lake Committee's leaders and lawyers, backed by twenty thousand members (along with the National Audubon Society and CalTrout), 16 years to build public concern and win battles in court that ultimately protected the lake and its tributary streams from the effects of stream diversions into the Los Angeles Aqueduct. The victory came in 1994, when the city's water licenses were amended to reduce diversions from Mono Lake and gradually raise its volume back to the level at which the unique saline ecosystem remains healthy, teeming with brine shrimp, alkali flies, and birds, and where dust storms blowing off the exposed lakebed stopped violating legal air-quality standards. Ubiquitous "Save Mono Lake" bumper stickers were replaced by ones saying "Long Live Mono Lake."

The key court decision relied on the public trust doctrine, which recognizes the government's legal responsibility to protect resources that have environmental and aesthetic value and that are therefore held in trust for public benefit. Applying that doctrine to Mono Lake altered water rights that had been held for a half-century by the City of Los Angeles, and set a legal precedent that water agencies facing similar issues elsewhere found unsettling. Yet the city was not devastated; instead, significant water conservation and recycling measures were put in place that annually save more water in Los Angeles than the city had been diverting from the lake.

We had witnessed an environmental victory that contrasts vividly with the Aral Sea devastation. In Turkmenistan, as river diversions to serve the distant city of Ashgabat and regional farms destroyed the ecosystem and subjected the region's population to harmful dust storms, no "Aral Sea Committee" formed to advocate for protecting the resource value of that massive inland sea. And no public trust legal tool is available to the people of eastern Turkey to protect Hasankeyf on the Tigris

Figure 24. East of the Sierra Nevada, limestone tufa towers are now exposed above the surface of salty, alkaline Mono Lake, where more than a million birds feed and nest.

River. Nor has there been a successful legal or public opinion campaign, as yet, to protect the diminishing Ogallala groundwater resource, perhaps because (as we heard in western Kansas) that problem is hidden from sight.

Geoff McQuilkin, the executive director of the Mono Lake Committee, discussed the lake's connections with water environments west of the Sierra. Reducing diversions from Mono Basin streams required Los Angeles to conserve water, which, Geoff noted, "affects how much is drawn out of the bay delta for Southern California and in turn affects users right there along your route. It's all linked together. We want to see good things happen in the Sacramento–San Joaquin Delta, where you are headed, because that means there will be less reason to come back and seek more water from Mono Lake." Geoff even saw connections tracing back through the millennia. "Millions of years ago this was part of the San Joaquin River drainage out to the ocean, before Mammoth

Mountain and the Sierra crest rose so high." The Sierra Nevada is still being forced upward by colliding tectonic plates. Near the end of our California trip we would cross the San Andreas Fault and leave the North American plate behind as we walked onto the Pacific plate at Point Reyes. Hydrology is a function of topography, and topography would be on our minds in the coming days as we faced the climb over the Sierra Nevada escarpment.

We took a deep breath, then walked out the door of our house a mile north of Mono Lake (38°02'N; 119°08'W) and headed west up Lundy Canyon, where Mill Creek gathers snowmelt from the surrounding peaks before flowing below our house into the lake. Lundy Canyon is full of aspen groves, dramatic wildflowers, and waterfalls. The canyon steadily angled southwest and, at the top, we stood exactly on the 38° line.

Connie Millar appeared out of the wilderness, moving at a brisk pace as the afternoon sun dropped toward the Sierra crest. Short and wiry, with a head of curly red hair, our neighbor is a U.S. Forest Service paleo-ecologist who explores connections between mountain vegetation and changing climate patterns. She told us that photographs taken 80 years ago documented that the nearby glaciers had shrunk by at least 80 percent during the past century. "North Peak Glacier may be gone in a decade," Connie said, "and all of the glaciers in the central Sierra in our lifetimes."

As we had seen in western China, some of the clearest signs of climate warming are showing up at high elevations. Scientists like Connie are trying to understand what global changes mean for local areas, because effects vary. "In parts of the eastern U.S., temperatures are actually cooling," she told us, "while the West is warming much faster than the global average." This kind of complexity confuses some people, who expect to correlate the big picture with what they see happening outside their homes.

A rising temperature trend has been recorded in California through the past 120 years, as the industrial age tapped into the carbon trapped in fossil fuels. But the slope of that curve has steepened over the past

30 years (coinciding with a doubling of global population that pushed human numbers beyond 7 billion). And "in the last 10 years everything has shot up," said Connie.

Forecasts give two-to-one odds that much of California will be drier by the end of the century—a 20 percent decline in precipitation—with more annual precipitation falling as rain and less as snow. A trend of earlier snowmelt runoff, shorter spring seasons, and longer summers is already apparent in the Sierra Nevada. These trends are similar to those projected for the western Mediterranean nations, which we heard about while in Portugal, and will complicate the timing of water storage and release by reservoir managers in both regions.

We were sitting near the tree line. Scattered clumps of trees consisted primarily of whitebark pine and some limber pine. Connie explained how "drier" added to "warmer" was killing trees. Higher "evaporative stress" is, as Connie put it, "pushing trees over thresholds they had not experienced in the last several hundred years, with hillsides going in a flash of mortality. But even where there is a lot of mortality, it's not total." Slower-growing pines were being taken out, producing strong selection for trees that may do better under the altered climate.

We all wondered what might happen to Mono Lake. It would be tragic if the lake's ecosystem collapsed because of climate change despite the "Save Mono Lake" victory. The inland sea survived prehistoric droughts, but the buffer taken away by 50 years of stream diversions has not yet been restored.

Connie's concerns were broader, stretching from Mono Lake to even the smallest riparian corridors. She has a special interest in "rock glaciers"—where ice lies embedded beneath insulating rock—something we could see at the lower edge of North Peak Glacier, hugging the mountain crest just south of us. The terminal moraine there was seeping water. Rock glaciers, with their insulating coats of rubble, should help wetlands stay lush and alive for many years during a warmer-climate regime. "They won't fill Mono Lake up," Connie explained, "but will serve these local wetlands and provide persistent streams where other canyons will just dry up. It's encouraging, not for the statewide water supply, but locally for the birds, and wildflowers and pika." American

Pika (*Ochotona princeps*) are close relatives of the Qinghai Lake species in China. The small rodents gather grass each summer to dry in "hay" piles for fodder during the long alpine winters.

California's official climate strategy is to prioritize efforts to conserve the most sensitive resources. "People are talking about triage these days," Connie said. "There will be things we may have to let go. On the west slope of the Sierra, almost all of the resource plans have a priority to re-introduce salmon, but by mid-century, waters may become too warm to support natural salmon runs, so you may not want to put the effort there if you are just going to lose in the long run."

It was a grim picture to consider as we sat in one of the world's most heavenly settings. There may still be time to modify this scenario, if political will can be mustered, here and abroad, to address the human pressures driving global changes in the atmosphere and oceans. International agreements have been ineffective so far, despite considerable renewable energy and low-carbon-emissions transportation measures we had seen in parts of Asia, Europe, and in our own country.

On a beautiful bluebird morning, we followed a trail up to Secret Lake. We remembered seeing mayflies hatching there in August, coming up like magic from the depths and appearing on the water surface, then popping their wings out to dry. The flies meant lots of protein for finches that snapped them off the lake. At the Sierra crest we paused to gaze westward across the wilderness lands in Yosemite National Park that we would travel in the coming days.

Grand Canyon of the Tuolumne
to Hetch Hetchy

Loaded with heavy backpacks, we slipped and slid down a rough trail to the McCabe Lakes, where we found fishnets stretched out across the water. The national park was two years into a five-year project to clean fish out and improve conditions for native Yellow-Legged Frogs. We saw no fish, but also no polliwogs or adult frogs, yet. It is easy to introduce fish into a lake, but very hard to get them out.

Sierra Nevada frogs are beleaguered not only by trout predation but also by a fungus that is infecting amphibians all over the world and has spread across much of our mountain wilderness, perhaps carried by flying insects. Some individual frogs will, we hope, have resistance, but as west winds coming off Central Valley farms blow in pesticides, which interfere with reproduction, the amphibians' prospects look bleak.

In the following days we descended through the breathtaking Grand

Canyon of the Tuolumne River, dropping past California, Le Conte, and Waterwheel Falls, gravity pulling the water, and us, down and down. We left the lodgepole forest behind and walked beneath black oaks and Jeffrey and sugar pines. The canyon featured a series of cascades and pools, with massive granite walls framing the views and channeling our direction of travel through a landscape that resembled Yosemite Valley, without tourists. We saw almost no other people. A long cascade into an emerald pool got us to stop for a dip. We had the place all to ourselves. In another cascade of pools a Dipper (Water Ouzel) was feeding underwater. It was a great place for both human and avian dippers.

Above 6,000 feet, it felt as if everything was in a hurry to set seeds before winter, including mountain ash, with its clusters of bright-red berries. Where we first encountered oak trees, acorns had been falling under canopies of yellow-brown leaves, but farther down the canyon the trees had barely begun to think about autumn.

The canyon opened wider at Pate Valley, where the elevation was only 4,350 feet (37°56′N). The air felt noticeably thicker and the trees were mostly oaks and incense cedar. Downstream from there the canyon curved south and then back west toward Hetch Hetchy Valley. No trails follow the upper shores of Hetch Hetchy Reservoir, so we instead faced a strenuous two-day detour. During congressional debates in 1909 over construction of a dam inside Yosemite National Park to serve as a San Francisco municipal water source, Secretary of the Interior James R. Garfield said, "The prime change will be that, instead of a beautiful but somewhat unusable 'meadow' floor, the valley will be a lake of rare beauty" (Egleston, 1909, 7). Plans were made for lodges and roads and trails, but once the dam was in place, the no-swimming-or-lakeshore-access policy was instituted so that city water would not have to be filtered. Or perhaps it has simply been in San Francisco's best interest to keep awareness low about changes their dam brought to Yosemite's twin valley.

We left summer behind in Pate Valley on the 3,500-foot climb to Harden Lake, and returned abruptly to early autumn at 7,600 feet (37°53′N). There were day-hikers there, and though we had 12 backcountry miles to do the next day, running into people made it clear we had finished the wildest part of our crossing.

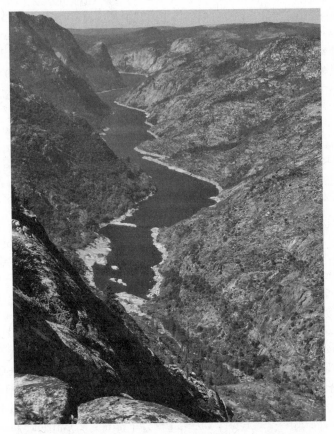

Figure 25.   Above Hetch Hetchy reservoir in the Yosemite
National Park wilderness.

Navigating switchbacks on the trail down toward Hetch Hetchy, we
glimpsed what struck us at first as a big slab of gray granite. Then our
wilderness-focused brains adjusted, and we realized it was the paved
road to O'Shaughnessy Dam.

In 1913, Congress authorized the dam that flooded Hetch Hetchy
Valley. About two-thirds the size of world-famous Yosemite Valley,
Hetch Hetchy is where the Tuolumne River once meandered through
flowery meadows and dramatic waterfalls plummeted off the cliffs
of towering granite domes. John Muir, leading the unsuccessful fight
against approval of the dam inside Yosemite National Park, wrote the

famous lines "Dam Hetch Hetchy! As well dam for water tanks the people's cathedrals and churches, for no holier temple has ever been consecrated by the heart of man" (Muir, 1912, 181).

In the backpacker's camp not far from the dam, Spreck Rosekrans found us. He was chairman of the board of Restore Hetch Hetchy and also handles water issues for the Environmental Defense Fund (EDF). He became our new best friend when we saw that Spreck was carrying fresh food and wine. Tall and fit, Spreck exudes contagious energy.

Restore Hetch Hetchy hopes to remove the reservoir from the valley, an idea that strikes many people as audacious, particularly the San Francisco Public Utilities Commission, given the water-supply challenges facing California. Yet feasibility studies, including one by the National Park Service and U.S. Bureau of Reclamation in 1988 and another by the California Department of Water Resources in 2006, concluded that existing storage downstream can feasibly replace the water and most of the hydroelectric power from Hetch Hetchy. But in the State of California's report, Spreck explained, "they reassessed cost numbers prepared by the Environmental Defense Fund [which estimated a $1.5 billion cost] and said we needed 30 percent more for engineering. They called for replacing the water by four times over! Their approach would keep a water bank in Don Pedro Reservoir, keep Cherry and Eleanor Reservoirs, then add another dam (in the coast range), plus groundwater storage, plus conservation, plus purchasing water from willing sellers. They called for all those things and concluded it would cost between $3 billion and $10 billion!

"Well, we said you could do one or the other or a mix. We want an academic, independent body to oversee these alternatives. The state said this study was as far as they were going to go and kicked any future action over to the federal government."

Restoration of Hetch Hetchy was given considerable federal support during President Reagan's administrations. "Don Hodel, secretary of interior under Reagan, made the suggestion in 1987 that restoration could occur with no harm to San Francisco," Spreck said. "Don is not someone you'd think of as a 'tree hugger,' but he understands the natural heritage here."

The Restore Hetch Hetchy organization formed in 1999 when it split off from the Sierra Club. It would have pleased John Muir to see this century-old battle re-engaged. Muir's last major conservation battle was over the dam. He died of pneumonia in 1914, the year after the dam's congressional authorization, but historians have speculated that Muir died of a broken heart over the loss.

"The damming of Hetch Hetchy was the event that turned the Sierra Club from an outings club to a political organization," Spreck said. "Two years after Congress authorized this dam, they passed the National Parks Act, basically ensuring that we're going to preserve parks and not do anything like building the Hetch Hetchy Dam ever again. The national park system might not exist except for what happened here at Hetch Hetchy."

For us, knowing that it is feasible to restore the sister valley to Yosemite without a loss of San Francisco's water supply, it seems a moral imperative to correct the hundred-year-old mistake—a gift worth pursuing for our children and the world. Damming and flooding a natural treasure like Hetch Hetchy was a bad decision, an error similar to the choice being made in Turkey today to flood the archaeological and historical treasures of Hasankeyf. In both cases, dam builders with shortsighted objectives seem oblivious to long-term values that matter to the entire world. Correcting the Hetch Hetchy mistake may take many years but remains feasible; there is only a small window of opportunity to avoid a similar mistake on the Tigris River in Turkey.

"One possible approach for our future effort," Spreck concluded as we cleaned up after the meal, "is a clause in the California constitution that says all water will be 'reasonably and beneficially used and methods of diversion shall be reasonable.' So," he asked, "is it reasonable to impound water into storage inside a national park?"

In the morning, Spreck walked with us across O'Shaughnessy Dam, and we looked at the exhibit panels. It seemed wrong that there were no photographs of the valley as it appeared before the dam was completed. Spreck did us another major favor by driving Janet and our backpacks to Cherry Lake while David walked the 15 miles, carrying just a daypack. Cherry Lake was the one part of the San Francisco water system that Spreck had never visited. Janet found a campsite and put signs up for our son Ryan, who was meeting us there with bicycles and supplies.

The trail that David followed (actually an overgrown road used to haul gravel to the dam site during construction) passed through oak and pine forest toward Eleanor Reservoir. That "lake" was nearly full, which always helps a reservoir's looks. Along with O'Shaughnessy Dam and the aqueduct facilities delivering water to San Francisco, the entire Hetch Hetchy system took 20 years to complete after the 1913 Raker Act. Water from Eleanor moves into Cherry Lake, which passes it down through hydroelectric generators and eventually into the Tuolumne River. Boating and the recreational uses typical in reservoirs *are* allowed in those two man-made lakes, unlike at Hetch Hetchy, because they do not supply drinking water to the city.

We reunited at Cherry Lake campground, at 5,000 feet (37°58'N). The mountains were behind us.

Friends of the River and
New Melones Reservoir

On the autumn equinox, we bicycled down through the Stanislaus National Forest, aware that all of us on the 38°N latitude line around the world saw the noon sun that day at 52° above the horizon (90° minus 38°). Our 40-mile ride to Sonora was an extremely hilly, forested, and scenic route. At one point we were pedaling so slowly uphill that a butterfly heading the same direction passed us by.

One of the canyons bottomed out at the Clavey River, an unusual water body in that it flows north to south, rather than east to west like most Sierra Nevada streams. Although there was graffiti on the bridge and an official "No Bungee Jumping" sign, this tributary of the Tuolumne River is actually one of the most pristine rivers in the Sierra Nevada. It flows dam-free for its entire length down to the Tuolumne and is one of only three Sierra streams without a major diversion or dam. The unique

Clavey Rainbow Trout lives in this state-designated "wild trout stream" and Forest Service "critical aquatic reserve," yet so far, the proposal to give the Clavey protection as a federal "wild and scenic river" has proven too controversial.

At Kurt Stegen's house in the foothills town of Sonora, we took our first showers in nine days, then Kurt drove us to the New Melones Reservoir visitor center. He is an arborist who had volunteered for Friends of the River for 35 years and personally experienced the contentious history of the Stanislaus River. Although at first he seemed quiet and serious, we soon realized Kurt's passionate resolve.

New Melones, completed in 1978, was the final dam built by the Army Corps of Engineers for the federal Central Valley Project. When first proposed back in 1944, it was intended simply to control floods, but Congress modified its authorization in 1962 by adding irrigation, hydro-power, fish and wildlife enhancement, recreation, and water quality to the list of projected benefits. Contention regarding New Melones was not about whether to build the dam, but how high it should be and how much of the river canyon it would flood. In 1974, Friends of the River sponsored a state ballot initiative, Proposition 17, to prevent the reservoir from inundating nine miles of a popular river-rafting stretch at the upper end of the canyon.

"I think the most incredible thing about Friends of the River was the grassroots campaign they put together," Kurt said. "There was no Internet, of course. I don't really understand how we communicated, but it was a well-organized campaign." In the 1970s, newsletters were the medium of choice—prepared on typewriters, using lots of correction fluid and making headlines with rub-on transfer letters, then photocopied before sending them out via snail mail (a term that could not exist until the Internet came along).

Another group, Friends of New Melones, formed to defeat the ballot measure, and confused voters with billboards that read: *Stop "Wild River" Hoax! Stop Pollution of the River!* That a new reservoir could release water in the summer to dilute downstream pollution justified the anti-pollution message. The "wild river hoax" claim was made because upstream dams controlled the river's natural flow. (Of course, many wild

Figure 26. New Melones Reservoir, on the Stanislaus River in the Sierra Nevada foothills.

rivers share that status, including the Colorado River through the Grand Canyon.) The *Los Angeles Times* editorialized that "the billboards seem an absolute betrayal of the truth to us. We too, oppose Proposition 17. But to call it a 'wild-river hoax' and to suggest that the proposition would result in pollution is a resort to tactics that have no place in responsible democratic campaigning" ("It's a Hoax, All Right," 1974, 6). Yet the tactic proved effective; Proposition 17 was defeated.

An exhibit at the visitor center summarized some of that history, which Kurt and so many others lived firsthand. From there we drove to the viewpoint overlooking the dam, a massive earthen plug 625 feet high, spanning 1,560 feet across the canyon. Kurt regularly takes school groups to the viewpoint, but a gate usually keeps the road closed, and we were the only people to walk in that day.

In 1979, as the reservoir began to fill, Mark Dubois, then the director of Friends of the River, chained himself to a boulder in a secret location that would be flooded. Several others joined that protest, forcing water to be released from the dam to avoid drowning the protesters.

Although the ballot initiative had been defeated, the Friends of the River campaign did produce considerable political support in Sacramento. In May 1979, Governor Jerry Brown vetoed a bill that would have allowed New Melones Reservoir to be filled to capacity. In 1980, the California State Water Resources Control Board set a low limit for the reservoir level, but heavy runoff after a record winter in 1982 trumped political will and filled the reservoir to its spilling point. In 1983, the board lifted its restrictions; a full reservoir was a fait accompli.

That record winter snowpack of 1982–83, coincidentally, followed the summer we arrived to begin work at Mono Lake. We had expected to watch the lake continue its long decline, but following that exceptional winter, it rose 6 feet, postponing the ultimate crisis point for the lake ecosystem. The dry stream channels feeding Mono Lake again ran with water that could not be handled by the Los Angeles Aqueduct system, but instead of defeating environmental goals, as happened at the New Melones Dam, the re-watered channels triggered efforts to keep that water flowing and fish alive in Mono Lake's tributary streams.

As for the long list of projected benefits from New Melones, it has not met expectations. A history of the project prepared in 1994 for the Bureau of Reclamation concluded that this particular dam could be considered "a case study of all that can go wrong with a project" (Simonds, 1994). There has not been sufficient water to meet all of the obligations used to justify the project.

In Korea, China, Turkey, Spain, and Portugal we had learned about the other side of the balance sheet, the long list of problems that dams sometimes create along with their benefits. Dams destroy natural processes of rivers. They trap sediments that otherwise build beaches where riparian vegetation and wildlife thrive, and they block salmon, steelhead, and other anadromous fish trying to reach ancestral spawning grounds from the sea. The benefits of flooding—scouring sediments, replenishing spawning gravels, and fertilizing floodplains—are lost. And so the Quercus group in Portugal called the Alqueva Dam a "white elephant."

Although Quercus and the Friends of the River were defeated in their respective campaigns, both organizations moved on to successes. Friends of the River campaigns helped protect many of California's major rivers, including the American, where construction of the Auburn Dam was stopped due to seismic risks. In the mid-1980s, the Friends founded the California Water Caucus, a statewide coalition to protect the Sacramento–San Joaquin Delta and promote environmentally sound water policy.

"I stumbled onto the Stanislaus River," Kurt said, "when I was 18 years old, through a job. In 1973, I didn't realize how involved with it I would become. I don't usually get caught up in things, but it had a special appeal to me." "Stumbling" into a commitment that ends up shaping a lifetime of concern is a pattern that other activists we met around the world would recognize.

Toward sunset, we stood at the upper end of the reservoir where the whitewater rafting stretch once began, and looked at tree snags poking above the water (the reservoir was only at 46 percent of capacity that autumn). For Kurt, that low level suggested an intriguing possibility: might the reservoir be managed to restore the whitewater rafting stretch, while still ensuring emergency flood storage capacity in high water episodes like those forecast to occur more frequently with global warming? Bringing back flowing water to the upper Stanislaus River may not be an impossible dream.

After we left the foothills town of Sonora, 12 miles of bicycling through the Gold Rush country brought us to Angels Camp. Samuel Clemens spent time in that little town in the 1860s (after first visiting Mono Lake) and heard a story in a saloon that inspired him to write "The Celebrated Jumping Frog of Calaveras County," his first story published with the byline Mark Twain. Angels Camp celebrates the annual Jumping Frog Jubilee each May. Regrettably, the California Red-Legged Frogs (*Rana draytonii*) that inspired the event are nearly extinct in the nearby foothills, partly because of predation by bullfrogs introduced from the East. Today's jumping competitors are all the much larger bullfrogs.

From Angels Camp we turned west again, dropping out of the foothills through Salt Springs Valley, a "lost" valley, off the usual travel routes. Ranch houses, cattle, and fields are scattered in a part of this crowded state seemingly bypassed by "progress." Just east of there, by contrast, retirement communities and golf courses sprawled across the lower foothills. We saw White Pelicans, Pied-Billed Grebes, and assorted ducks in the reservoir at the lower end of the valley. The road we were traveling was lined with gravel tailing piles left by miners who had seemingly chewed their way across the land; these historic remnants reminded us of similar present-day piles on the Jade River in Hotan, China.

The hard physical part of this trek was over once we reached Stockton, since we would proceed by boat in the morning. Hiking and biking had given us a new, physical appreciation for the size of California.

The Sacramento–San Joaquin Delta

Perched on the dock with our gear, we watched a parade of boaters come and go: mom, pop, dog, then several lone fishermen and five head-shaved, tattooed young men crammed into a boat built for speed rather than fishing. John Knott's 33-foot Catalina sailboat emerged from among a group of college rowing-crew sculls. It was to be our mother ship for the next five days.

An experienced sailor in those waters, John had been the district superintendent for Sierra State Parks, which included responsibility for Mono Lake. Now retired, he had brought along David Martin as crew and sailed from San Francisco Bay to meet us. John's brother, Marty, sailed along beside us in his smaller catamaran. It was sunny, hot, and amazingly clear. That afternoon we boated only a few miles in the Stockton deepwater shipping channel. Shortly after leaving the dock, we intersected one of the proposed routes for a controversial peripheral canal that could reduce the volume of water flowing through natural delta channels even more by redirecting it around the periphery of the estuary to the southern aqueducts.

The Sacramento–San Joaquin Delta gathers the waters of California's two major river arteries into the largest estuary on the U.S. West Coast, with 1,000 miles of channels that flow toward San Francisco Bay and the Pacific Ocean. About 75 percent of California's water demand from cities, suburbs, and farms originates south of this point, while about 75 percent of the precipitation that falls in the state occurs to the north. To create twenty-first-century California, water must be shifted from north to south in aqueducts served by massive pumps at the south edge of the delta. Populations of fish—salmon, smelt, and shad—that depend on natural delta flows have plummeted as water diversions have altered the ecosystem, especially since exports increased from a yearly average of 4.6 million acre-feet, to 6 million acre-feet between 2000 and 2007. The damage to the estuary ecosystem and the competing thirsts of cities, commercial fisheries, and farms make the delta not just a key transfer point but a hub for the state's never-ending political and environmental battles over water.

Toward sunset that first night on the delta, we turned out of the deep-water channel and anchored near Lost Isle. David Martin had a portable GPS unit; he paddled his kayak a few yards away, then called back that he was *exactly* on the 38th.

It was a beautiful evening in a natural setting. Swimming in the warm water felt fantastic, though we could not help speculating about the agricultural chemicals that were in the runoff mix. A biologist with the U.S. Fish and Wildlife Service has called the lower San Joaquin River "the lower colon of the valley's river system" because it carries so much pollution (Carle, 2003, 145).

With the sky glowing orange behind Mount Diablo's silhouette, we also tried out the kayaks that had been leashed to the back of the sailboat. One side of the channel was green and lush and noisy with birds. The other side was a bare rock levee, devoid of life (part of 1,600 miles of levees lining the delta channels). As night fell, from behind that levee we heard the insistent beeping of vehicles backing up. The next morning we saw the source of the noise: trucks driving atop the levee, each pulling two trailers brimming with tomatoes harvested during the night.

Mount Diablo had been visible ever since Stockton, and would be with us all the way into San Francisco Bay. The peak is an important landmark for California geographers. Although it does not sit atop an intersection of exact latitude and longitude lines (37°51'N; 121°55'W), Diablo's 3,845-foot summit looms high above other coast range hills. On the clearest days, the Sierra Nevada and Mount Shasta are visible from there. That explains why, in 1851, Colonel Leander Ransom, a surveyor with the General Land Office, chose Mount Diablo's peak as an initial point. He set a marker on the summit and then used it to survey prime meridian and baselines. Working outward from those lines, surveyors divided California and Nevada public lands into townships (squares with six-mile sides). The perpendicular angularity of many of the Central Valley's roads and fence lines reflected those imaginary lines running across the landscape, originating from Mount Diablo.

Life on the river is a whole different world, its pace dictated by tide, wind, and marina. It was hard to believe that the little channel in front of us could take us to the Golden Gate and the open sea, but as we chugged out the gradually widening channel, the scale of the delta revealed itself. We do not have many big rivers in California, but the delta gives a taste of the Mississippi and Ohio Rivers. We could feel the power of the water, full of life, giving life, moving life. It had a wonderful smell, like "sweet rain," Janet decided; an altogether different feel and odor than a Sierra stream.

On our first full day on the river, we saw the dark head, then the thrashing body, of a Sea Lion and the flash of a huge silver fish. It was an exciting glimpse of a sight now rare in the delta as salmon numbers have declined. Conditions were very different a half-century ago when the rivers teemed with salmon passing on multiple runs far up into the mountain waters to spawn and add their bodies to the river food webs. California has four distinct runs of Chinook Salmon (*Oncorhynchus tshawytscha*), named according to when they pass through the delta upon returning from the sea. The winter run species is endangered, the spring run is threatened, and the two fall runs are "species of concern," according the U.S. Fish and Wildlife Service. Millions of Chinook used to swim upriver each year; in the winter of 1883–84, more than 700,000 were caught for delta canneries. Dams constructed in the 1940s, '50s, and '60s dealt a severe blow by blocking access to the upper reaches of Sierra Nevada rivers and streams; only 10 percent of suitable spawning habitat now remains. Less water flowing out to sea and carrying the chemical signatures of each salmon's birth water has brought the species to the brink.

Midday, we paused near a flooded expanse within the delta known as Frank's Tract for more exploring in the kayaks. Sixteen miles straight south from there, pumps serving the California Aqueduct sent water toward Southern California cities, and others supplied the Delta-Mendota Canal for farms in the San Joaquin Valley. If we had let the boat drift with the current, we might have been pulled toward the pumps instead of downstream, because their suction is powerful enough to reverse flows in delta channels.

We had been given an update on issues affecting the delta when we were at Hetch Hetchy, as Spreck Rosekrans had just attended a Bay Delta Conservation Plan (BDCP) meeting. The BDCP group, appointed by the governor, was seeking agreement among water diverters, environmental interests, and delta farmers and residents—a process Spreck characterized as "gnarly." Water agencies want the peripheral canal built either around the delta or in tunnels beneath it, so they can continue moving freshwater if levees fail and saltwater reaches their pumps. Many of the 2,500 delta residents worry that, with a canal in place, the other interests would no longer share the "common pool" that gives everyone an incentive to maintain levees. Major concerns remain that enough flow be guaranteed to the estuary ecosystem and that a peripheral canal not simply facilitate ever more diversions.

One canal option envisions two massive 33-foot diameter tunnels beneath the delta that might not be full except during the wettest, surplus-water years, but would be capable of shifting large volumes to southern state reservoirs. That much capacity, however, was worrisome: California's development history makes it hard to trust loosely worded assurances that sufficient water would run out to San Francisco Bay at all times.

We sailed past several restoration projects trying to bring back some of the original tules; the effort looked tiny against the scale of acreage that has been cleared, plowed, drained, and controlled. For a short while that afternoon there was enough wind to put up the sails, though the sailors grumbled throughout our cruise that the calm weather was *too* fine. Our second night on the boat was at the marina at Brannon Island (38°07′N).

John Cain met us there the next morning. He was the director of restoration programs for the Natural Heritage Institute. (In the 1990s John worked for the Mono Lake Committee.) John's dark eyes lit with enthusiasm as he described his work. The Natural Heritage Institute protects and restores aquatic ecosystems and the services they provide to humans. Our time with John had to be early in the day because he had to attend an afternoon BDCP meeting in Sacramento as cochairman of a habitat restoration committee.

While we toured in John's car, the boats went on downstream. Our road ran beside the Sacramento River where massive wind turbines decorated the hills beyond the channel. "This channel of the Sacramento River is artificial," John said. "Ships used to get stuck in the Gold Rush days, so they dredged here and dumped the deposits along the edge. Now the deposits are the only soil that can be taken for rebuilding islands or strengthening levees around the perimeter of islands."

Farm soils on delta islands subside once the peat soil, naturally underwater in the marshes, is exposed to air. Oxidation converts complex carbon compounds into carbon dioxide gas and causes one to two inches of lost soil depth each year. So most delta islands are today below sea level, protected by surrounding levees and serviced by groundwater pumps that constantly run to keep fields dry enough to grow crops. Electricity to run the pumps can be a farmer's biggest expense.

On Twitchell Island, 12 feet below sea level, native tules and cattails were growing on several acres of ponds. John reached into several inches of water and came up with a handful of dripping black mud. "The rate of tule muck accumulation is about two inches per year. We're reversing subsidence—actually building up this island. These subsiding islands had become a substantial source of atmospheric CO_2, so this converts a carbon source to a carbon sink. We're hoping to repeat this on a much larger scale."

We mulled over "carbon-capture farming" as we drove to the south end of the island to see an experimental "green levee" covered with lush growth. Engineers prefer that levees be bare so they can see signs of failure, but during a recent flood, John said, "when people thought the standard levee was going to go, everyone took shelter behind this levee armored with vegetation."

Levee failure on an island in the north part of the delta could send salt all the way to the aqueduct pumps at the south end. "When a big gulp of saltwater suddenly moves in," John explained, "it can lead to a chronic problem. Not all of the salt on a breached island leaves when the tide goes out. That salt can be dissolved back into the water and gradually shifted by high tides farther and farther toward the aqueduct pumps." Saltwater, of course, is useless to farmers or for drinking water, so intact levees are critical.

We crossed Sherman Island, 11,000 acres that were theoretically protected by levees at the western margin of the delta, but the walls were built on sand foundations likely to undergo liquefaction during a large earthquake. Across the Antioch Bridge, the rural character north of the river channel was replaced by shopping malls, housing tracts, and crowded highways in the communities of Antioch and Oakley.

On a dairy farm near Dutch Slough, east of Oakley, construction of 4,500 homes had been planned, but the farmer instead sold his 1,200 acres to the state. The land's elevation—six feet above sea level (high ground in that region)—made it suitable for the largest freshwater tidal marsh restoration project in the delta. We walked atop a levee trying to visualize the dairy pasture as tule wetlands. "Come back in 10 or 20 years," John told us, "and you should see valley oaks and sycamores and native walnut trees grading out into grasslands and tidal marsh."

The restored wetlands, intermittently inundated by high tides, will provide important salmon habitat. Salmon spawn and are born much farther upstream in the mountains, but juvenile salmon, beginning their own treks from the Sierra Nevada to the sea, take advantage of tule marshes to pause, feed, and grow. They fare better later when they finally enter the ocean as larger salmon.

Our final destination with John was a subdivision built below sea level in Oakley, one of the epicenters of the home foreclosure crisis in Northern California. The levee to protect the houses met 100-year federal standards, so homeowners were not required to buy flood insurance. With climate models forecasting a possible four-foot sea-level rise by the end of the century, such faith in levees to protect houses seemed a recipe for disaster. China's Yellow River provides lessons about the potential for disaster for people living below water level behind breakable levees.

John delivered us to our mother ship, waiting near the confluence of the Sacramento and San Joaquin Rivers. The channel grew wider and wider as we sailed through Suisun Bay and, just before sunset, docked in Benicia, the major nineteenth-century commercial port for produce shipped from Central Valley farms to Bay Area cities, until railroads took over that job.

Strait to the Bay

The following morning, the fourteenth day of our cross-California trek, the extremely low tide inside the Benicia marina left even the shallow-draft catamaran temporarily aground. The bay's surface was nearly flat, the weather sunny and calm—perfect for paddling the 9.5 miles to Vallejo in kayaks. That morning will remain in memory as one of the highlights of our trek.

As we approached the Carquinez Strait, several massive ships came in beneath the bridges and generated impressive wakes that rocked our little boats even though we hugged the shore. Cars looked tiny on the freeway spans of the bridge high overhead. Paddling beneath them, we whooped out loud with exhilaration. We had arrived in San Francisco Bay, along with vital flowing water that had originated high in the Sierra Nevada.

On the far side of Carquinez Strait, we rounded a breakwater and entered Mare Strait. At that point the rising tide began pushing us up the channel, and we barely had to paddle to reach the Vallejo Municipal Marina, where we waited for the sailboats to catch up. There we took Marc Holmes on board and continued up the Napa River estuary.

Marc, tall and tanned, with dark curly hair, is the bay restoration program director for the Bay Institute. He guided us a few miles up the Napa River to former Cargill salt ponds now being restored (38°09′N). Of the original 190,000 acres of tidal marshes from Suisun Marsh out through San Francisco Bay, 79 percent are gone. But opportunities for coastal wetlands restoration in the Napa-Sonoma marshes are enormous.

While we motored upriver, Marc pointed toward houses lining the northeast shoreline. "None of those residential areas could be permitted today, because they're built on the tidal margins of the bay." The San Francisco Bay Conservation and Development Commission was established in 1965 to halt infill development, because by then, one-third of San Francisco Bay had been filled or diked or drained. That followed a successful campaign by Save the Bay. An older organization than the Bay Institute, Save the Bay functions much like the Chesapeake Bay Foundation to educate and advocate for this West Coast mirror image. It is a striking coincidence that the nation has two gigantic coastal bays intersecting the 38th parallel.

Today, any flatland one sees bordering the water "used to be bay," Marc said, "including San Francisco's financial district and Mission Bay." Communities along the edge had expanded by simply filling in shallow marshlands. "Plus, there were forty-eight garbage dumps on San Francisco Bay in marshes."

"Like 'Mt. Trashmore' in Alameda, where I grew up," Janet said. "I remember going with my dad over there. You just dumped whatever."

"Sure. Bring appliances down, refrigerators, back up your station wagon, and watch a bulldozer push the trash out into the marsh."

Everyone transferred to Marty's catamaran to motor into the salt pond complex. Tidal marsh there was converted to grazing land in the mid-1870s and later to ponds for duck-hunting clubs. The shallow wetlands were perfect for phenomenal numbers of migratory shorebirds and waterfowl. In 1956, Leslie Salt bought the properties and later sold them to Cargill. We were looking at salt evaporation ponds that each represented about 1,000 acres of former wetlands. In 1994 the ponds were bought by the state and were now managed by the Coastal Conservancy.

"This complex worked by bringing bay water in through a channel and circulating it through a series of twelve ponds," Marc said. "Eventually it was as concentrated as it would be. Decant off the brine solution, and scoop up the salt crystals left sitting on these flats."

"Where did the levee go?" we wanted to know. Only a broken line of low vegetated clumps separated us from the nearest pond.

"Eighteen months ago restoration crews made strategic breaches and razed them with bulldozers." Using aerial photographs and old maps to select breach locations that matched old channels, the workers excavated the channels inside the ponds before cutting the breaches.

"Look out there! I'm stunned to see all that cordgrass." Marc was clearly excited at the pace of change. "This grass is brand new, in the last 18 months, without any planting; seeds traveled in on the tides. None had grown here in 100 years or more. This will start the natural cycle of a tidal marsh once again: detritus will fall, the elevation will gradually build up, and eventually we'll recover large pickleweed marshes."

We asked Marc to take us through the rest of the developing food chain. "Now you've got the grass coming back," we said—"what else?"

"First are invertebrates; they come back immediately. Then wading birds: Black-Necked Stilts [*Himantopus mexicanus*] and American Avocets [*Recurvirostra americana*], and sometimes some short-legged waders as well, when the tide goes out. With inundation at high tide come Long-Billed Curlews [*Numenius americanus*], some dabbling ducks, occasionally some diving ducks like Canvasbacks. In the wintertime, these ponds are just lined shoulder to shoulder with migrating birds.

"This is a very important fish nursery," Mark added, where many fish that spend their adult lives at sea come to reproduce. That is a key function of coastal estuaries everywhere, including the Yellow River Delta and the *marismas* of the Guadalquivir River in Spain. "It is also the farthest-west habitat for Delta Smelt, historically." The endangered Delta Smelt species is at the center of controversy over water diversions because those powerful aqueduct pumps suck in and kill so many of the tiny fish.

"This is so great; you're seeing it right at the beginning," Marc said. "Come back here in five years, and the cordgrass lines will extend across the pond, and you'll have to get into kayaks to paddle up the channels."

Salt pond restoration was also under way in the south bay, complicated by mercury in the bottom mud that had washed down from mines before the ponds were diked and, with restoration, may be released into the bay. Mercury becomes concentrated as larger fish eat smaller fish, leading to health warnings about humans eating too many fish caught in the delta and bay. This explained a sign on Brannon Island that had startled us on reading the first two lines:

Eating Fish Caught in the Delta
May Cause Pregnancy
or Slow Development Problems

As at Chesapeake Bay, pollution has been a major problem for the West Coast bay, culminating in the unrestrained runoff of sewage and industrial wastes that poured directly into the water during and after World War II. The national Clean Water Act of 1977 provided the tool for controlling such "pipe" sources, and since then, the odor and healthfulness of the bay have improved tremendously. Today, as in the Chesapeake and Ohio River watersheds, the problem is how to control nonpoint pollution from millions of diffuse sources. Whenever it rains, storm drains bring a wave of petrochemicals and toxins washed off streets and yards. "Drains to the Bay" messages stenciled above sidewalk storm drains are as ubiquitous here as they were in Virginia.

Our goal the next day was to reach China Camp State Park and the neighboring national estuarine reserve. "At China Camp, you'll see the marsh as it existed 4,000 years ago when the bay was formed," Marc told us. "Petaluma marsh is the largest tidal marsh remnant. China Camp's is smaller, but one of the wonderful things there is a complete cross section of the ecotone from open-water bay to mudflat to tidal marsh to grassland to oak woodland. It's a precious little remnant."

A party atmosphere developed that afternoon—perfect weather, good company, beer, peanuts, and conversation about good work being done to restore the bay. After we dropped Marc off at the marina, the tide and weather were right for crossing, so we sailed over to the Petaluma marsh that Marc had described. It was nice to turn the motor noise off when Captain John put up the sails for a while.

Near the mouth of the Petaluma River, a few feet beyond a crumbling bank, the Sonoma Baylands marsh spread eastward, part of the San Pablo Bay National Wildlife Refuge. Just upriver was the start of the Petaluma marsh. After much anchoring, rope tying, maneuvering, and discussions about currents, we settled in for the night at 38°07′N.

In the morning we sailed south across San Pablo Bay toward China Camp. From the middle of the bay, we could look back eastward toward the Carquinez Bridge and, looking southwest, see the distant skyscrapers of San Francisco. The catamaran had to be used for the shallow-water landing, so we said goodbye to Captain John and David Martin. On the ride up onto the beach, we felt like Columbus coming ashore in the New World. China Camp beach is exactly on our latitude line. While we waited for our son Ryan to arrive, we checked out the little museum and cafe.

China Camp's only resident, Frank Quan, is a spry, elderly man who runs a small café. In the late nineteenth century, about five hundred Chinese people, including his family, lived in this shrimp camp, one of many around the bay. Shrimp were sent all the way to China, as there was no local market in California. Two million pounds of shrimp were harvested each year until the Chinese bag nets were banned as a shrimping technique (mirroring the harvesting-versus-sustainability conflicts in the Mediterranean Sea). As with fishing of historically over-harvested salmon whose reduced, yet still viable populations had persisted in the delta until recent years, Frank had continued to fish for shrimp in the bay. Now the species is almost all gone because shrimp seek out the mixing zone to find less salty water, and water diversions from the delta mean less dilution within the bay. Historically, during the high flows of spring runoff, most of San Francisco Bay would be freshened. Frank no longer expected the shrimp population to last through his lifetime. The shrimp cocktail he served us in his restaurant came from Oregon.

On the bicycles again, thanks to Ryan, we continued across the Marin Peninsula through redwood groves, coastal scrub, and pastures where cattle grazed. At Olema, where we reached State Highway 1, the ground was displaced as much as 21 feet on either side of the San Andreas Fault by the earthquake of April 18, 1906, more infamous for its devastating effects on the City of San Francisco. Tomales Bay, a long arm of water a few miles to the north, is where the fault enters the ocean. By lunchtime we had left the North American tectonic plate for the Pacific plate and were at the Bear Valley headquarters of Point Reyes National Seashore (38°02′N).

Park superintendent Don Neubacher had attended the graduate School of Ecology at the University of California, Davis, with two of the founders of the Mono Lake Committee. He spread a map on a picnic table outside the park office and told us about the many projects the Point Reyes staff were working on to restore coastal marshes and reopen miles of streams to fish passage. Their biggest single effort was restoration of the 560-acre Giacomini marsh at the south end of Tomales Bay.

"It had been a dairy," Don said, "and they diked off the bay in the 1940s. Lagunitas Creek flows through there and is the one creek in this region that still has a pretty good run of Coho Salmon, with 15 to 20 percent of the remaining Coho in central California. You'll see three-foot fish in 12 inches of water."

Don told us that the Giacomini project alone adds 10 percent to central coast wetlands. "That gives you an idea how little is left," he said. "On October 25, we'll open the last bit of levee, and 50 to 60 percent of the land will flood at high tide. This will be transformed in a couple of years."

Camping that night in a volunteer campground not far from the headquarters, we heard the screechy calls of Spotted Owls from the nearby trees.

Point Reyes and the Pacific Coast

The morning was overcast as we pedaled up the west shore of Tomales Bay, but we found the sunshine as we turned back south toward Drakes Beach. Our son Ryan had driven there already and was photographing California Gulls (*Larus californicus*) on the beach when we arrived. We wondered if they were gulls born at Mono Lake—some of the 70 percent of that species in California that fly to the lake each spring to breed and raise chicks. During the winter those birds spread along the entire state coastline, but these had perhaps chosen the most direct migratory route, following the 38th parallel as we had, to reach the ocean.

We inflated our two-person kayak and changed into wetsuits and life jackets. The bright-yellow boat, our "rubber ducky," though lacking the grace of the wooden kayaks we had relied on in the delta, had to carry us only two miles across the bay.

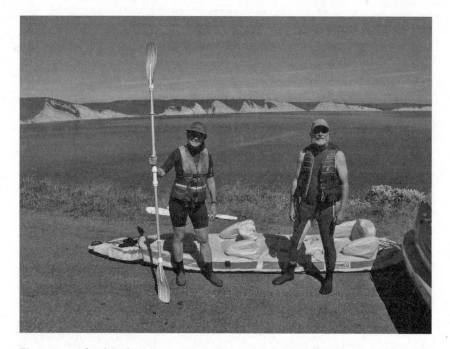

Figure 27.   After kayaking across Drake's Bay, the authors walked across Point Reyes to the lighthouse where the 38th parallel reaches the Pacific Coast.

We were novices at launching through waves. Three aborted tries sent us tumbling about on the beach, but Janet looked down the shore and said, "The waves look easier there." We carried the boat 100 yards, studied the wave pattern, then made a successful launch.

About halfway to the fish dock at Chimney Rock, we heard the wonderfully mysterious calls of loons and then saw several of them dive and surface, calling repeatedly. Then from the beach came the guttural, vibrating "chonk" call of Elephant Seals (*Mirounga angustirostris*). A half-dozen cigar-shaped seals were on the beach. Two reared up, displaying their protuberant noses to each other.

Ryan met us at the fish dock and we loaded gear into the car. Then we finished our trek across California by walking across Point Reyes.

The three miles still to go lay almost exactly along the 38th parallel. Cows studied us solemnly as we passed their pasture. A turnout provided a view straight over a cliff onto another gathering of Elephant Seals. We strolled up the road, feeling curiously reluctant about finishing the walk. At the Point Reyes lighthouse parking lot, friends were waiting with food and champagne. Together, we walked to the lighthouse over-look (38°00'N; 123°01'W).

We stared at the ocean, trying to realize that there was no place left to walk. The 220-mile span across our home state had seemed vast when walked at two miles per hour, sailed at four miles per hour, and bicycled, at times, up steep hills at a "speed" easily exceeded by a cruising butter-fly. During those seventeen days we had actually meandered 350 miles, 75 of them on foot, 168 on bicycles, and 107 miles on boats.

Off to the southwest, the profiles of the Farallones Islands were vis-ible. Over the horizon, far away, the 38°N latitude was heading toward another landfall in Japan. Soon we would travel that direction.

For now, we took turns proposing toasts and speaking about the trek. Although it was another gloriously sunny, clear day, a foghorn blew in the background. Point Reyes is the "windiest place on the Pacific Coast and the second foggiest place on the North American continent," accord-ing to the National Park Service. Before we left the overlook, an elderly couple who had been coming there for more than 30 years, told us that weather conditions that day were the best they had ever seen at that spot. We were very lucky.

After finishing the cross-California trek, we soon returned to the Golden Gate to join migratory bird counters there, to embark on a whale-watching cruise to the Farallones Islands, and to see the first high-tide inundation of the Giacomini marsh at Point Reyes—fitting points of clo-sure for the California exploration.

Hawks, Merlins (*Falco columbarius*), and Turkey Vultures (*Cathartes aura*) soared above the Marin headlands north of the Golden Gate Bridge as we learned about the Golden Gate Raptor Observatory from its direc-tor, Allen Fish, and volunteer hawk counter Herb Brandt. Both had been with the program for more than 20 years. Raptors migrating between North and South America seek out short water crossings like the one at the Golden Gate, at the mouth of San Francisco Bay (37°49'N). Just as in

Sicily, at the Strait of Messina, and at the mouth of Chesapeake Bay, birds are concentrated, and most easily seen and counted, where water and land funnel their annual migrations away from frozen winter conditions in the autumn and back when spring reawakens life in the north.

The day before, 390 raptors had been tallied from Hawk Hill. Southward-traveling migrants arrive in late September, coming down the north coast and from inland California, some heading all the way to South America. Spring migrants return between March and May.

The Hawk Hill site was once part of the military bunker complex built to guard the Golden Gate. The land opened to the public in 1972 after the Golden Gate National Recreation Area was founded. The raptor count began after a "crazy lady," Allen said, "began calling the California Academy every fall, saying she was seeing a Broad-Winged Hawk [*Buteo platypterus*] every autumn by her house above San Francisco. No one had ever seen a Broad-Winged Hawk here; it is an eastern woodland hawk that migrates down to the Amazon." As it turned out, the lady was not crazy at all; there were over one hundred Broad-Wings every year migrating with many other types of raptors. "It's the only spot in California to predictably see this species," Allen said. "Nobody had a clue that they were moving through here before that lady called."

"Here's an accipiter right in front of us," Allen said, "a Sharpie [Sharp-Shinned Hawk]. Look how quickly he's getting lift." The birds gain altitude on thermals rising over the south-facing hillside. "The 'money of migration' is lift," Allen added.

Herb, a highly experienced volunteer, explained that "birds climb really high here, and then go out over the water and start dropping. They'll go out sometimes and come back, go higher, and finally take a line across."

The raptor observatory is funded by the Golden Gate National Parks Conservancy and has three hundred volunteers working in the hawk count program. Allen was proud of the conservancy "for seeing there were ways of bringing people into a national park to do conservation work who didn't have to be PhDs, didn't need to be professional contractors; that they could create a community tradition."

"The 'ah ha!' moment occurred for me about five years into this," Allen said. "The truth is, this program offers therapy. There's a quality of daily calming and easing into urban life. People feel like they have a meaningful, useful, long-term connection to the landscape. The joke is, 'Take two oak trees and call me in the morning.' What I feel like I'm 'selling' is more spiritual than scientific. The state of your inner nature is directly parallel to the outer nature where you spend time. That's why this is so much bigger than just monitoring migrating birds."

Our whale-watching boat chugged out of sunny San Francisco Bay through the Golden Gate and into a dense marine fog that persisted the entire 28 miles to the Farallones, the westernmost bits of California on our 38th parallel route (37°44'N; 123°02'W). A marine biology student, Jessie Beck, was with us to get her first close look at the islands before returning for 12 weeks of research there for the Point Reyes Bird Observatory.

After three hours' traversing ocean swells, we finally approached the jagged "Devils Teeth" rocks looming through the mist. A lighthouse that stands on a small flat area of the 70-acre island was once staffed by light keepers and members of the Coast Guard who lived in two houses down below; now the light is automated. Today researchers occupy the houses, staying three to six months to study seabirds, marine mammals, and sharks. Because there is no freshwater available on the islands, other than a brackish spring contaminated with guano, and because outside water is not only expensive to transport by boat but also very difficult to move up the cliffs from the dock, captured rainwater is funneled off roof gutters into a tank. That water is filtered multiple times before researchers drink it.

Nesting bird colonies seek out islands because the surrounding water protects them from land-based predators, but during California's mid-nineteenth-century Gold Rush, people plundered the Farallones for eggs. By 1900 they also wiped out Elephant and Fur Seals, but now all wildlife is protected within the Gulf of Farallones National Marine Sanctuary. More than 300,000 breeding seabirds use the islands—the largest colony in the lower forty-eight states. Only a few scientists are today allowed to set foot on the islands.

Jessie's interest in the Farallones began when she was young and her family took trips to Point Reyes. The islands "seemed mystical and impossibly remote," she said. "My parents always asked us if we could see them in the distance, and we would strain our eyes looking." Occasionally the fog would lift enough for glimpses of the island peaks, just at the horizon.

Having spent 12 weeks studying Farallones seabirds, Jessie later told us that "living on the islands was, first and foremost, long days of hard work, seven days a week. My life became consumed by the writhing mass of birdlife. I woke up to the calls of Western Gulls, knew each nest along each pathway, spent the days monitoring those nests, ate dinner while watching baby Common Murres [*Uria aalge*] fledging, and fell asleep to Cassin's Auklets [*Ptychoramphus aleuticus*] chirping. I began to recognize each bird's personality, their distinct calls and gestures to their partners. In the midst of a screamingly loud, guano-splattered chaos, I began to see the dense colonies as beautifully simpler microcosms of our own societies. It was the most rewarding work I'd ever done."

During our boat cruise, we saw sea lions and Humpbacked Whales cruising near the islands, but none of the Great White Sharks that are common there in other seasons. The boat trip itself was challenging enough to leave us in awe of researchers like Jessie who commit to harsh living conditions and isolation on the Farallones to better understand the needs of recovering species.

About five hundred people were at Point Reyes to help Don Neubacher and the national seashore staff celebrate the rewatering of the Giacomini marsh. Led by a ranger, we walked across the valley to meet the tidewater, then steadily backed away as the leading edge kept pushing onshore. Humans were not the only spectators enjoying the event; rising water was evicting rodents from underground burrows, and a host of egrets and White-Tailed Kites (*Elanus leucurus*) were patrolling the edge of the tide. Though this was just the beginning of a process that will restore the complex wetland ecosystem, it was exciting to see pastureland "going tidal" (as T-shirts on sale that day proclaimed) after so many years of planning, fund-raising, and preparatory work.

After the California coast, our next landfall along the 38th parallel would be in Japan, another place undergoing renewal, and the final destination we needed to reach before reflecting on our completed travels around the world.

PART FIVE

Renewal and Recovery

"The toki [Japanese Crested Ibis] has the power to change people."

Ryugo Watanabe, general manager of the Sado City Biodiversity Promotion Section (Fatakuchi, 2010)

Map 5. Japan

Recovery on the Tsunami Coast

Tsunami coming. Must go up!

Advice to an American tourist in Matsushima, Japan, from a local woman, on March 11, 2011, minutes after the magnitude 9.0 Tohoku earthquake (Paul, 2011)

On Honshu, Japan's largest island, our original plan was to explore the mountains-to-sea watersheds and the relationship between wetlands and rice farming—the biocultural landscape the Japanese created on their islands. Floodplains cover less than 10 percent of the land in Japan, but hold 51 percent of the population and even more of the nation's economic resources. The risk contained in those living sites became evident after a massive earthquake occurred off the northeast coast of Honshu. Its epicenter was at 38°19′N.

As we flew into Tokyo, the map displayed on the airplane-seat screen depicted not only landforms and our plane's location, but also physical features along the ocean bottom, including the Japan Trench, a dark-blue gash extending north to south off the east coast of Honshu, Japan's largest island. That deep trench appears where the Pacific plate bends downward and is overridden by continental plates. This collision zone on the Earth's crust explains why Japan is one of the world's most earthquake-prone nations.

The magnitude 9.0 quake on the morning of March 11, 2011, the fourth-largest ever recorded, created a tsunami that reached the island's east coast in 30 minutes. A 30-foot wave struck Sendai, the largest city near the epicenter (38°15'N). Farther north, in Miyako, the wave funneled up a narrow coastal canyon that elevated its crest to an astounding 132 feet.

The tsunami caused much more damage than the quake itself or its many strong aftershocks. Almost twenty thousand people died and over half a million were displaced. The quake also damaged nuclear reactors south of the line, most critically at Fukushima Daiichi, where several reactors lost cooling capability, experienced meltdowns, and released radiation into the air and ocean. We arrived in Japan six months later, after major recovery operations, food and water shortages, electricity outages, and the threat of radiation had been addressed.

Japan is a land of contrasts: quietly speeding bullet trains, carefully tended gardens, clean well-kept cities, but also bright flashing lights, karaoke bars, and pinball machine gambling halls. It is a traveler's dream country of order, efficiency, and politeness. The chaos brought by the quake and tsunami were incredibly upsetting for such an orderly people.

Our flight arrived in Tokyo, and, before leaving there to head north to the 38th parallel, we visited the Ministry of Environment's national headquarters to learn about a plan for a new national park on the coastline affected by the tsunami. Several existing national, prefectural, and local parks will be incorporated into the Sanriku Fukko ("Recovery") National Park. *Sanriku* is the name given to the northeast coast of the Tohoku region (a designation something like "New England"), which includes four prefectures (equivalent to our states).

The humidity and heat of central Tokyo reminded us of similar weather we had experienced in Korea. At the Nature Conservation Bureau offices, we located Shinjiro Sasaki, an assistant director of national parks. The office space in a large room was a warren of side-by-side desks, dimly lit by the windows and the lights of computer screens, because overhead lights were turned off. The air conditioning was also off and everyone wore short-sleeved shirts. That summer, the national government had asked the nation to take such energy-saving measures to reduce the likelihood of power blackouts, because all but fifteen of the nation's fifty-four nuclear reactors were shut down after the earthquake. Mr. Shinjiro gathered several other administrators and planners into a grand meeting room with large windows overlooking downtown Tokyo (we were on the twenty-sixth floor). The director of national parks, Tetsuro Uesugi, and the director general of the Nature Conservation Bureau, Tsuneo Watanabe, joined the meeting.

Over tea, they told us how plans were progressing for the new national park. A vision planning meeting had occurred just four days earlier, but much work remained to begin the project, including forging contacts with the communities struck by the tsunami.

"What prompted this idea for a new park?" Janet asked.

"At first, we were overwhelmed by the earthquake disaster," our translator summarized, after Mr. Watanabe and Mr. Sasaki answered. "We were here in Tokyo, but wanted to help those people because the national parks there are a key to their local economy and the most significant features in much of that area. We have staff in several parks there now, but hope to expand with this new concept for a national park." As we would see later that week, Japan's national park rangers implement national conservation policy, not only on park lands but also in communities and regions where environmental projects are under way. This project will increase tourism, provide employment to locals as park guides, and honor the victims of the tsunami. Damaged facilities will be preserved, and boulders that the tsunami carried inland will be left in place so visitors can see the giant wave's impact. A new trail will be built along 200 miles of the coast, to not only connect the parks but also provide escape routes to higher ground during future tsunamis. The park service will also aid the local fishing industry by promoting ecotourism on fishing boats.

Funding was still an unresolved issue—they all chuckled and rolled their eyes when asked about the project budget—but Director General Watanabe said that special reconstruction funds established after the quake by the legislature might be available, as well as private funding. He emphasized that the Sanriku Fukko National Park, though still only at the vision stage, was now his agency's most important effort. Beach cleanup of tsunami debris had begun that summer, directed by the Ministry of the Environment.

We would visit Matsushima Bay, in the southern portion of the proposed new park, where the nation's first "natural park" was created in 1931, renowned as one of Japan's three most beautiful sites. Natural parks are operated by local prefectures rather than the national parks division; one challenge for the Tokyo office will be coordinating with other levels of government affected by the new park project.

The Tohoku bullet train from Tokyo to Sendai was so smooth and silent, it was hard to believe we were going 200 miles per hour. A scheduled stop was Fukushima City, 39 miles inland from the Daiichi nuclear power plants on the coast. Although the city lies well outside the 12-mile mandatory evacuation zone and is separated from the coast by a low mountain range, some radiation had drifted that far. Efforts continued there to scrub roofs and remove radioactive materials from private houses, parklands, and meeting venues, with a particular emphasis devoted to households with children—all aimed at reassuring local citizens about conditions in that city.

The tsunami wave that struck the Daiichi nuclear facility on March 11 was 46 feet tall when it swept over a 20-foot seawall meant to protect the power plants. The meltdown of three reactors that resulted was eventually declared a "Level 7 event," the highest on the International Nuclear Event Scale (equivalent to the Chernobyl meltdown in 1986). The mandatory 12-mile evacuation zone imposed by the Japanese government affected more than seven thousand businesses and more than one hundred thousand residents. Some of those unfortunate people were still living in shelters six months later, while the situation at the nuclear reactors themselves remained extremely serious. There was plutonium pollution at the power plants, and radioactive water still leaked into the Pacific Ocean.

We were glad that stops are brief on the Shinkansen bullet trains. Ours soon sped away from Fukushima City toward Miyagi Prefecture's capital of Sendai, sixty miles farther north. Within a half-hour of the March 11 earthquake, a 30-foot-high tsunami wave had swept across Sendai's airport, carrying cars, trucks, and tons of debris inland. Videos of that dramatic scene, with people gathered on the airport terminal roof as the wave swept inexorably onshore, were televised again and again during news coverage of the earthquake. In Natori, the airport city, the tsunami killed 1,019 people and destroyed or damaged more than 8,000 homes.

The tsunami did not touch central Sendai, which is on higher ground and had only minor quake damage, but for weeks afterward, more than 1 million people had no electricity or gas and were isolated by damage to roads, railroad tracks, and the airport. Air service resumed a month later and trains began running on April 29. After six months, hotels for tourists and business travelers were open, and life was more normal; tourists were trickling back.

Sendai is west of Matsushima Bay at the south end of the Tohoku region. A ten-mile train ride from downtown brought us to the local port for fishing vessels and tour boats. Twenty-two people had died near the port, a relatively small number because the force and height of the tsunami were abated by offshore islands. There are 260 small islands in Matsushima Bay, most of them crowned by pine trees (*matsu* means "pine," and *shima* "island"). The first tsunami destruction we saw was a damaged seawall and sidewalk along the waterfront. The woman who sold us our tour boat tickets came outside to show us where the wave had crested four feet up the wall of the terminal building as she and others escaped to the roof.

Before boarding the boat, we bought one of the souvenir T-shirts for sale everywhere in the region that proclaim *"Ganbarou Tohoku!"* (Let's Do Our Best, Tohoku!) or *"Ganbatte Nippon!"* (You Can Do It, Japan!).

We had the morning tour to ourselves for the 50-minute trip, which wound among the scenic islands and finished at the town of Matsushima. The famous bay remains a magical place, worthy of its reputation for beauty. The larger islands are inhabited by fishermen and seaweed farmers. A small one is named "Horse Set Free Island," because samurai used to turn old horses loose there to live out their days. Most of the islands are even smaller rock stacks, intricately sculpted by the water, topped by elegant-looking Japanese red pines (*Pinus densiflora*), a symbol of longevity in Japan. Here and there, trees had been broken by the force of the tsunami or killed by the saltwater inundation. Several wrecked boats still lay on their sides on beaches, six months after the earthquake.

At the town of Matsushima, population 16,000, we located the headquarters of Matsushima Bay Natural Park. The office building is only a few feet from the bayshore and, though badly damaged by the tsunami, had since been rebuilt. No one spoke English in the small office, but one of the park administrators, Hiroya Miura, kindly led us on foot for several blocks to the tourist information kiosk at the train station, where a woman could translate. Mr. Hiroya was aware of the new Recovery Park being proposed in Tokyo.

When asked his opinion of that effort, after a long pause he said, "It will be difficult to coordinate so many different levels of national and local park operations." "But," he added, "yes, it could be a way to increase tourism and provide jobs." The national park planners in Tokyo have much work to do.

A bridge to one of the closest islands offshore, just behind park headquarters, had been completely destroyed by the tsunami and not yet rebuilt. On the September day we visited, waterfront cafés bustled with Japanese tourists having lunch. The cafés were part of a line of businesses that is separated from the bay only by a two-lane road and a low seawall. The ground floors of all those buildings were flooded when the relatively small tsunami wave arrived (only about five feet high). That left a mess to clean up, but did relatively little structural damage. The shaking itself caused more damage in this village, destroying two hundred homes, including our English translator's (it had already been rebuilt). Snow had been falling when the quake hit, and power was knocked out, so keeping warm at night was added to the many challenges residents and travelers faced right after the quake.

A long graceful bridge over to Fukuurajima Island survived, though damaged tiles on the approach walkway were being repaired. A chorus of strange frog calls provided the background serenade to our hour-long hike around the lush island. Labels identified many of the three hundred species of trees and grasses found there. Coves and beaches reached via

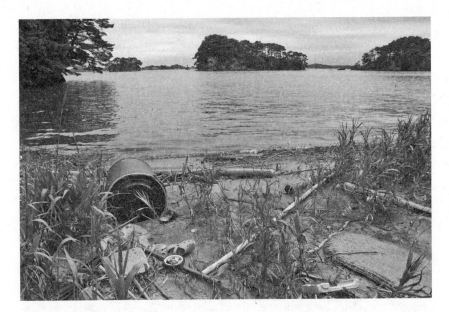

Figure 28. Debris piled by the March 11, 2011, tsunami includes children's shoes, in a cove off Matsushima Bay.

the main trail appeared pristine, but a little-used side path dropped down to a remote cove still covered by tsunami debris. Among the pieces of lumber and assorted flotsam were a thermos, lightbulbs, a wheel from a child's scooter, and two shoes of different colors and sizes, leaving us with unanswerable questions and a sudden sense of the losses that had occurred.

This region has a long history of both earthquakes and tsunamis. Sendai was swept by a massive tsunami 1,142 years ago following an earthquake of unknown size, according to written records and layers of sand deposited in inland soils. In 1896 the Meiji-Sanriku earthquake, magnitude 7.2, generated a tsunami that destroyed about nine thousand homes and killed at least twenty-two thousand people farther north along the Sanriku coast. During the twentieth century, no earthquakes reached magnitude 8.0.

As the country rebuilds, the question is whether to relocate villages to higher ground or to concede to the desire of thousands of evacuees who, willing to gamble that the next tsunami will not happen for many generations, want their old lives back where they grew up and worked. Official government policy is not to make that long-term mistake; relocation is nonnegotiable, according to the national Reconstruction Agency and city governments that aim to build tsunami-resistant communities.

Two months after the big earthquake, the U.S. Geological Survey had counted 525 aftershocks exceeding magnitude 5.0, including five above magnitude 7.0. On July 31, 2011, the region felt a 6.4-magnitude aftershock, at a time when the disaster-ravaged population had just endured a torrential summer rainstorm that dumped more than 26 inches in three days and led to flooding-induced evacuations in the Fukushima and Niigata Prefectures.

During the night after we visited Matsushima, an earthquake shook our beds in Sendai, waking us up near midnight. It was only magnitude 4.7—a minor temblor, hardly noticeable in that part of the world.

We departed that morning on another train, heading west again along the 38th parallel through the island's interior. Although the local park official foresaw many obstacles to the plan being considered at national park headquarters, in the next few days we would see an impressive collaboration among farmers, rangers, scientists, and citizens aimed at adjusting a way of life in order to rescue the endangered Japanese Crested Ibis. If such determination and energy represent the national norm, recovery and renewal along the Sanriku coast will almost certainly also be achieved.

It Takes a Village to Save the *Toki*

Halfway across the volcanic interior of Honshu Island, we switched from a train to a bus because heavy rains that summer had damaged tracks along our route to the west coast city of Niigata (37°55′N). With Japan's dense population and limited space, the forested mountains of the interior contrasted with land otherwise intensely cultivated or inhabited. Historically, water development in Japan has been closely tied to cultivation of rice in paddies, mostly situated on converted wetlands. The island land base is relatively small, and its rivers are short, descending steeply from mountain snowpacks to the sea. Even though the nation today has more than 2,600 dams, "their combined storage capacity is less than that of Hoover Dam in the United States alone, and only six percent of the total storage capacity of the dams under [federal] control . . . [in] the United States" (Water Environment Department, 2012).

Japan's rapid economic growth after World War II, achieved as it bene-fited from the demographic effect of a lower national birth rate, coincided with occasional regional water shortages and increasingly severe pollu-tion. In the 1950s, cadmium pollution from mines along the Jinzu River led to *itai-itai* (it hurts, it hurts) disease, so called because patients' bones grew so brittle that any contact caused them to cry out. Mercury contam-ination in the Agano River, which flows from Fukushima Prefecture to Niigata, led to the nation's second large outbreak of Minamata disease, named for the town in southern Honshu where six thousand residents had earlier come down with severe mercury poisoning. Industrial pol-lution, city wastewater, and nitrate runoff from farms caused frequent red tides in the Seto Inland Sea (surrounded by Honshu, Shikoku, and Kyushu Islands), where phytoplankton levels ballooned, killing fish in offshore "dead zones" (similar to those in Chesapeake Bay). Since the late 1960s, these problems have prompted the passage of pollution con-trol laws that have had some success, but national water pollution issues remain serious, according to the Ministry of the Environment.

From the train and bus windows, we saw ubiquitous rice fields, turn-ing golden yellow just before the harvest. After World War II, changes in centuries-old methods of rice farming introduced pesticides and weed-control herbicides, along with concrete channelization of irriga-tion canals. Those changes destroyed the food chain that had fed the Japanese Crested Ibis, and led to the species' extinction in Japan. Rice paddies were strikingly beautiful where they met the edges of dark-green forests. That border zone between mountain foothills and ara-ble land is *satoyama*, the habitat type critical for the ibis, which is called *toki* by the Japanese. The large pale-pink bird is the avian symbol of the nation; its scientific genus and species names reflect that special signifi-cance: *Nipponia nippon* (*Nippon* means "Japan").

At the port city of Niigata, serving the west coast of Honshu Island, we would catch a ferry at dawn to Sado Island. Professor Tsuneo Sekijima of Niigata University met us the night we arrived in the city, and intro-duced us to the *toki*'s complex and compelling story.

Once common across Japan, the ibis thrived for thousands of years wherever rice paddies came to dominate much of the landscape. With wingspans up to 4.5 feet and a very long down-curved bill, *toki* have a

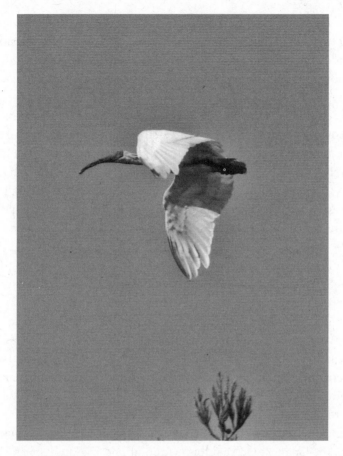

Figure 29.   Japanese Crested Ibis, called *toki,* fly again on Sado
Island thanks to a breeding program that brought them back
from the edge of extinction.

stunning appearance. In the past, they roosted in trees close to the paddies where they fed, the same locations where people lived and worked,
and the birds became a familiar, precious part of national life. Although
similar in size to a Great Egret, Crested Ibis have shorter legs and need
to forage in water less than four inches deep. In the mud and shallow
water of rice paddies, irrigation canals, and natural wetlands, the sensitive tips of their long beaks probe by touch for insects, worms, and fish.

There had always been hunting pressure on the birds, shot for food and for their beautiful pink feathers, but the decline in the late twentieth century coincided with modernization of rice farming. Besides the herbicides and pesticides that destroyed their food, the concrete-lined irrigation channels, with new gates to control floods and allow farmers to drain their fields, physically kept fish out of the paddies, especially a slim loach that was the birds' key prey. Loach could no longer swim up from flowing channels to spawn. To facilitate the use of heavy farm tractors (in place of livestock), rice fields were dried up before the harvest, which shut down biological productivity each year. Trees the *toki* needed for roosting and nesting had been cut for fuel, especially during the two world wars. By 1971, Sado Island (bisected by the 38th parallel) became the only remaining place where *toki* flew in the wild in Japan. They had also been disappearing across their extended range outside Japan, including eastern Russia, eastern China, and the Korean Peninsula.

In 1981, in a desperate effort to save the species, Japan's last five wild *toki* were caught, and captive breeding efforts began. That same year, a remnant population—only seven birds—was rediscovered in China, where the species had also been considered extinct. The Chinese began a successful captive breeding program and, in 1999, gave Japan two *toki*, then sent one more in 2000 and two more in 2007. The last of Japan's once wild birds died of old age in 2003.

There is, clearly, no point in breeding and releasing birds back into the same environment that caused their extinction in the first place. The story of the Japanese Crested Ibis is unique, and was an inspiring way to end our around-the-world travels, because the urban and rural residents of Sado are embracing key changes in their island lives and livelihoods. Dr. Sekijima's research to assess the habitat needs of *toki* is just one part of an unusual cooperation among university, government, environmental organizations, and local people, including children and, critically, the certified ecofarmers of Sado who now produce "Toki Brand" rice.

We were on the first ferry from Niigata, leaving at dawn for the two-hour trip to Ryotsu, the port city of Sado Island (38°05'N). A ranger wearing the yellow uniform shirt of the Japanese national park system waited there to meet us. The large island (over 200,000 acres) is mountainous, with its tallest peak at 3,845 feet, and is home to sixty-two thousand people, of whom about 25 percent are involved in rice farming.

We spent the rest of that first day with national park ranger Kei Osada. Gray-streaked hair belied his young appearance and energy. In a 15-year career as a ranger, he has worked elsewhere, including the Sanriku coast and at the Tokyo headquarters, but had come to Sado Island just 14 months earlier to work at the Crested Ibis Conservation Center. "The role of a ranger in Japan is changing," he told us. "Now we are encouraged to be involved in community conservation efforts as well as taking care of parkland." The day he had mapped out for us was the perfect illustration of that new role.

Our first stop was a community work project where local farmers, members of the Biodiversity Conservation Network of Niigata, Sado City officials, and park staff had gathered to remove Large-Mouth Bass (*Micropterus salmoides*) and Blue-Gill Perch (*Lepomis macrochirus*) from a farmer's pond. Introduced 30 years ago from the United States for sports fishing, the nonnative species devour loach, so critical to the *toki* diet. The farmer who owned the pond, Mr. Zutsumi, was present, and we met a Sado City resident who cooks at a noodle restaurant, one of a dozen men and women volunteers. The pond had been drained, so bass and perch were visible flopping around in shallow pools. Under the eyes of local TV cameras and reporters, the fish were snagged with nets on long poles, exciting a flurry of thrashing fish and flying mud as men in waders heaped them into boats and rafts. The thousands caught that day would be distributed to local people to eat. Ranger Osada was in the middle of the action, scooping up fish, then being interviewed for local television.

Signs along the edge of the surrounding rice paddies showed that the farms met requirements for Toki Brand rice. To receive Sado City's certification as an ecofarm, a rice farmer must adopt methods that nurture living things, such as dirt-lined irrigation ditches (rather than concrete), more suitable for loach and insect habitat; establishing small wetlands called "biotopes" close by that are rich in food for the endangered birds; building fish ladders or ramps to connect ditches with paddies; and flooding fields in the winter, after the rice is harvested. Ecofarmers must

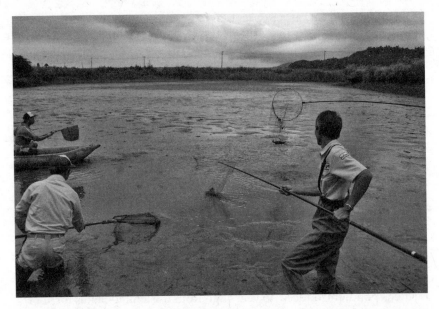

Figure 30. To help ensure the native food supply for the Crested Ibis, rice farmers, environmentalists, Sado City officials, and national park rangers work together to remove nonnative fish.

also use levels of pesticides and chemical fertilizers at least 50 percent below those typical for that region, and report the results of annual biological surveys to the city.

Certified farmers then can market Toki Brand rice in bags bearing a prominent picture of the pink bird. Their product is priced higher than that grown elsewhere, but the whole nation is aware of these efforts to save the species, and the premium rice sells well in major Japanese supermarket chains. Until the program was established and Toki Brand rice proved itself marketable, many farmers were predominantly negative about the Crested Ibis project, saying that "the release of Crested Ibises will fail" or that the "effort [was] irrelevant to agriculture" (Sekijima et al., 2010, 140). But after the first release of birds, in 2008, the number of participating ecofarms tripled to 695 by 2010.

The city's ecocertification guidelines rely on research findings by Tsuneo Sekijima and his cadre of graduate students, and also on the work of Yoichi Kawaguchi, a river restoration specialist from Tokushima University. He had joined us for dinner in Niigata and, on our second day on Sado, showed us field examples of his research. His studies clarified the irrigation system effects on *toki* and their prey and the benefits of biotopes, and he designed the fish ladders being used.

Cleaning out every single fish in the farm pond was going to take all day. Meanwhile, Ranger Osada drove us to the Ishidu River where a group of twenty elementary-school children who had enrolled with Sado City's environmental education program on a wildlife survey were catching river creatures. Although the channel was lined with concrete, the water was still rich with small fish, insects, and even fuzzy-clawed crabs, which the kids captured in clear plastic containers to be identified, drawn on art paper, and finally returned to the channel. This was part of the local effort to improve conditions for ibis related to "the small things" and to involve *all* members of the community in that effort. Sado City also has a Kids' Paddy Project, where farm kids and children from urban areas grow rice in pesticide-free rice fields, cultivate biotope habitats, and study the response of living things.

Kei then drove us up a winding mountain road as the sky grew darker and a few raindrops began to fall. On Takano Takeshi's farm, built on terraces in a canyon skirted by oak, beech, and pine trees, we joined twelve students from Wasada University who were energetically stirring mud in a rice paddy to enhance oxygen flow in that biotope. They had volunteered to come from their homes near Tokyo for four days; the concern for *toki* is national and felt in places other than Sado Island. Mr. Takeshi is the son of a local farmer who became famous for hand-raising the offspring of Japan's last wild *toki*, after it was born in captivity. One of the earliest memories he can recall is of *toki* perched in the trees outside his home. Ever since, he told us, they have remained special to him.

We had yet to see an actual Crested Ibis, though all of these projects were ultimately intended to benefit the birds. After a snack with the college students that included "*toki* milk" in a carton shaped like a fat bird (even Sado Island's dairy farmers are involved), Ranger Osada took us to the conservation center, where he spends most of his days, and we finally got a close look at Japanese Crested Ibis. As we came through a gate in a high fence, Kei raised his voice and spoke to the birds so they would not be startled. The cage held eighteen ibis that were moved to this training cage from eight other rearing and breeding cages. In the big training cage, they can fly, socialize, and get ready for their transition to the wild. Each bird eats forty loaches a day (raised at a fish farm), plus special pellet meal supplements. Food slides down to feeding ponds inside the cage from a nearby building, so that human contact is minimized. All the cages are monitored with video cameras serving a bank of screens in Kei's outer office (as at the center for Iberian Lynx we had visited in Spain). Despite the security systems, in March 2010 a marten (a member of the weasel family) broke into a cage and killed eight birds. After that setback, Ranger Osada transferred from Tokyo to Sado Island to help enlarge the center's staff.

About two hundred thousand tourists come each year to the Toki Forest Park visitor center, outside the rearing area, to learn about and see birds in cages. That operation is partially supported by the United Nations GIAHS, or Globally Important Agricultural Heritage System.

The conservation center's first release of birds came in 2008. For the first time in 27 years, ten *toki* flew wild over Japan. That day, fifth- and sixth-graders from Gouya Elementary School in Sado City sang the school's song for the Crested Ibis, to "encourage the birds to fly high in the sky." Marking the significance of that moment for the nation, the Japanese imperial family was represented by Prince and Princess Akishinomiya. Although the first releases were from individual boxes, in each year since, a less disturbing, "soft release" technique has been used: simply opening the doors of the training cage and letting the birds decide when they are ready to leave. In 2010 some birds took five days to depart.

Another nineteen *toki* were released in 2009; thirteen took flight in 2010 (the year of the marten attack); and Kei would release eighteen two weeks after our 2011 visit, including eleven males aged one to five and seven females aged one to three. The timing, late in September, coincides with the end of rice harvest and is when the wild birds naturally gather into bigger flocks for the winter. The goal is for sixty *toki* to be flying wild by 2015.

The relatively small wild population is backed up by 132 *toki* that now live in the conservation center (plus 41 more in three zoos in Japan).

Although there have been three earlier releases, and some of those free birds have paired up, built nests, and laid eggs, none of those eggs successfully hatched.

China now has more than 1,800 *Nipponia nippon*, with about 200 in the wild. Reproduction there has been more successful, perhaps because the Chinese began their breeding program with wild birds, instead of birds raised in captivity. Japan hopes someday to add more birds from China, in part to widen the genetic bottleneck that occurs when a whole population descends from just a few individuals.

Our busy day continued into the evening, as Ranger Osada escorted us to a dinner with the Wasada University students and local residents. The students enthusiastically practiced their English, telling us about their four-day adventure on Sado. They were curious about why we were there, asked what we had seen elsewhere on the 38th parallel, and wondered if our book would be available to them in Japan once it was published. Some of them, not surprisingly, were majoring in biology or environmental studies, but several were economics, sociology, or psychology majors. Perhaps, after that volunteer experience, they will find ways to incorporate an environmental focus into those university majors. The evening culminated with a performance of traditional Sado Island dancers and drummers.

Seeing such a rare bird up close, albeit in cages, had been a pleasure, but early next morning, we saw wild *toki* flying free. That was an exceptional thrill.

Our escort was Chihiro Endo, a soft-spoken post-doctorate student who looked too young to already have her PhD. The *toki*'s morning flight from groves of trees where they roost overnight is fairly predictable. Right on schedule, as the sun rose above the tree line, a flock of six birds suddenly appeared. We saw several other flocks at a distance, heading for feeding grounds, and then three birds flew low over our van, before settling in trees, which gave us a great look through Dr. Endo's spotting scope.

"*Toki* need a certain landscape of rich rice fields with trees alongside to rest and preen in," Dr. Endo explained, reinforcing the importance of *satoyama*. "Just because there is a lot of food available in a place does not mean they will come." In fact, when rice is close to maturity, as it was during our visit, the birds cannot forage among the tall stalks.

In the midday heat, the birds perch high in trees to rest and preen. With no flight activity to watch, Dr. Endo switched to surveys of their prey—insects, frogs, and loach—in a fallow rice field. Unexpectedly, a single *toki* flew over, delighting us all.

One of the volunteer bird surveyors that morning was a local carpenter who found a couple of *toki* feathers in the paddy (whitish at that time of year, but with a pink shaft). He also showed us footprints made by the *toki* in the mud and another set made by a Japanese raccoon. That gifted amateur naturalist gave us a set of excellent *toki* photographs he had taken. We asked if we might use one in our book, giving him credit; he gave us permission, but preferred anonymity.

Sado Islanders have committed to making the changes in their human system necessary to sustain the special birds that inhabit their island. The mayor of Sado City, Koichiro Takano, is an enthusiastic promoter of this effort. (Kei said that people consider their mayor close to being a *toki* himself, he is so involved.) Mayor Takano describes the effort as "a direct response to the biodiversity crisis that confronts humanity. Our efforts to bring back the *toki* through nurturing the small creatures on which it thrives is an attempt to restore the environment that is worthy of world-wide attention" (Takano, 2010, 7).

On September 27, 2011, two weeks after our visit, the door of the conservation center's training cage was opened at dawn. Within three hours, four male and three female ibises had flown away. Once the more cautious birds joined the others, fifty-one Japanese Crested Ibis flew, once again, over the island where the species had once gone extinct.

The recovering population of Japanese Crested Ibis is a symbol of hope and renewal and a model of sustainability for Japan and the world. The people of that small island, supported by the rest of the nation, are showing that environmental renewal can revitalize an economy, while that improved economy, in turn, helps fund restoration efforts. The endangered birds require a healthy water environment, as do the people of Sado Island.

As do the people of the world.

We Are Bodies of Water

What is a line but a dream and the will to follow it? What is
an adventure but a challenge to oneself to go a little further
off the beaten path?

From Janet's journal

Circling the globe along the 38th parallel is a 19,600-mile journey. On
the ground in Europe, Asia, and North America, the latitude line tra-
verses about 10,800 miles; the Pacific Ocean is 5,300 miles across, and the
Atlantic about 3,500 miles.

China and the United States both span similar distances along the
38° line. The longitude is 119°E at the Yellow River Delta on China's east
coast; at the western edge of that country, having crossed 44 degrees of
latitude, we were at 75°E. It is striking that the longitude numbers for
the United States are nearly identical, though measured in the opposite
direction from the zero point of the Greenwich Prime Meridian: Mono
Lake is at 119°W (the Pacific Coast, at 123°W), and the Atlantic Coast at
Assateague Island is 75°W. Move away from the latitude line to the north
or south in either nation, and national boundaries no longer coincide

so closely; this is an intriguing 38th parallel phenomenon. Both nations have moist eastern zones and arid western deserts. The climate and landscape are different because America's western border is an ocean, while China borders the mountainous continental interior of Central Asia.

These facts answer a question many people asked us: was the weather the same everywhere along the latitude line? It was not, of course. Nevada's desert is very different from the prairie in Kansas, which is unlike the Missouri woodlands or the green pastures of Kentucky. It was helpful to personally experience the transition from humid East to dry West in the United States, and actually feel the abrupt change near the 100th Meridian, where hair stopped curling due to eastern humidity and Janet suddenly needed lip balm again.

When we reached the Berry farm in Kentucky, Tanya, Wendell's wife, speculated that a familiar orientation to the sky and angle of light might make people comfortable elsewhere along their "home latitude." That particular relationship to the sky, which changes when traveling north or south, can become very noticeable if made in a fast jump by air. Holding to our 38th parallel course, we found that different altitudes and distances from the ocean instead produced the most noticeable changes in landscape and weather.

Summer rains that arrive with the Asian monsoon, like those we experienced in Korea and saw evidence of in the landscapes across China and Japan, were the most striking weather contrast with California's climate, whose Mediterranean pattern confines most precipitation to the winter. Korea's and Japan's summertime humidity can be overpowering to travelers from the western United States. The climate was much more familiar in western Turkey, Greece, Italy, Spain, and Portugal.

The Asian countries of Japan, Korea, China, Turkmenistan, and Turkey also took us furthest from familiar cultures and languages. Reaching the Korean DMZ, we met high school students who aim to shape the future of that environment, which is recovering, ironically, behind barbed-wire barriers between warring states. Yet it was distressing to watch South Korea's national river "restoration" project destroy riparian wetlands.

Cranes fly where China's Yellow River empties into the sea on the 38th parallel, but as we crossed that nation and witnessed its burgeoning economy, the new "holy crane" of China appeared to be construction cranes whose gangly towers and "beaks" were at work almost everywhere. New coal plants are rapidly being built in China, yet the nation also shows laudable concern for the climate-change problem they aggravate, by aggressively pursuing alternative energies and water-saving procedures for processing coal.

China's decision to build dams on its southern rivers and move water to the north to serve the thirst of growing cities prompted us to ask the same questions we have asked about Los Angeles and that Glennon Zelch asked about Las Vegas: why not decide the ultimate size of a city according to its available regional water resources? When will growth stabilize, as it inevitably must? And what quality of life and environment will humans choose, if they give considered thought to the long term?

Many of the water problems we encountered remain unresolved. In Turkmenistan, no government agency or organization seems devoted to effectively reducing river diversions to help the Aral Sea. Although international voices speak out against Turkey's plan to flood ancient Hasankeyf, that government still continues to build the Ilisu Dam on the Tigris River. Tensions between Turkey and nations downstream that share the Tigris and Euphrates heighten concerns about peace in the Middle East. The environment of the Mediterranean Sea seemed "tired," with too few birds and sea mammals, as if worn out by centuries of human exploitation. Intensive olive culture in Spain and Portugal demands more irrigation water in nations already pushed to the limits for water. Despite decades of effort to clean up Chesapeake Bay and restore oyster and crab populations, growing human numbers outpace those efforts. Mountaintop removal coal mines still push mining debris into the creeks of West Virginia and Kentucky. The Ogallala Aquifer, which supports so much grain and cattle production in the center of the United States, is overdrawn. Plains farmers keep pumping water out of the ground faster than it can be replenished and sending it around the world as corn syrup and beef products, but their lifestyle and economy must inevitably change as pumps go dry. Las Vegas maneuvers to acquire the groundwater of northern Nevada to serve a never-ending thirst for growth.

And yet . . .

We found inspiration in the work of Choony Kim and her co-workers with KFEM in South Korea. We were amazed by the dedication of Ran Liping and Green Camel Bell in western China, and stand in awe of Anna Giordano's brave, successful efforts to change a Sicilian culture that used to slaughter migrating hawks as they congregated at the narrow water crossing from Messina to mainland Italy. Spain's Iberian Lynx breeding and recovery program in the Guadalquivir River watershed is impressive, as is the uphill battle for the environment by the Quercus organization in Portugal. Environmental educators of the Chesapeake Bay Foundation must feel satisfying hope at the end of each day during which they give schoolchildren from the upper watershed their first experiences with crabs and oysters.

Despite considerable effort to plan our days, along the way we had unexpected encounters with people eager to share their lives and stories. Conversations often began, "Where are you from?" After we explained how and why we came to that place, stories would emerge. If we got little personal information, often we heard about someone else we should talk to.

In that serendipitous way, we met the county clerk on the west bank of the Mississippi River who kept a little museum open and eagerly talked about his former life as a riverboat pilot. There was the man outside Louisville, Kentucky, who had a passion for local history and who kept an amazing collection of old hand tools and other artifacts at a local history museum that we found only because a heavy rainstorm forced us off the main highway in search of shelter.

Passionate people are the rays of hope shining around the world.

We learned about Leo Dry, who established the Pioneer Forest in Missouri. Believing there was a better way to grow hardwood forests, he set out to demonstrate his theories on his own land, changing the accepted paradigm for forestry on watersheds in that region.

In the San Luis Valley of Colorado, Paul Robertson with the Nature Conservancy sat us down in a café, said, "This is how the story began . . ." and ran through the entire groundwater protection efforts of that valley.

Mark Rude, with the Kansas Groundwater Management District No. 3, works to protect the Ogallala Aquifer near Garden City. He was a fantastic source of information—a thoughtful man who could see and explain the big picture.

So many good people were doing good work along the 38th parallel. Those people were unanticipated discoveries during travels that, when we first conceived this project, focused more on exploring the water landscapes along our route. Many of the people—such as the staff of the Pioneer Forest or the local activist and hydrologists working to clean up uranium tailings in Moab, Utah—had been at their locations for many years and had no plans to move away in search of greener pastures. They recognized the value of settling in, building their knowledge of the land, and caring for the local environment and their local society. Each of these people nurtured a passion about place. We could see it in their eyes; we will always remember those young foresters in the Pioneer Forest and how proud they were.

Many Americans relocate too often to build such strong feelings of responsibility for their homes (despite the lifework of poet and writer Wendell Barry, which eloquently extols the importance of place). Jokes about environmental activists who travel the world to tell people to stay home miss the point. Traveling with one's eyes open and mind engaged broadens knowledge and attitudes, but a long-term commitment to home fosters a depth of knowledge and caring that is too easily lost in a transient society.

One of the most striking memories from these travels was the mountaintop removal coal story. We both had a visceral reaction to seeing the ripped-apart mountain beside Larry Gibson's place. We could not appreciate the tragedy of mountaintop removal until we saw that hole at our feet and the walls of debris choking the stream channels. We cannot believe that anybody thinks that is a good idea. For people living there, like Larry, the loss is impossible to imagine.

For more than 20 years Larry beat the drum about MTR, and never gave up, before his sudden death by heart attack in 2012. He did not just throw up his hands, but fought back. It became his life's work. He was a knowledgeable and eloquent speaker who came across so well, partly because he was a common man. In expressing his concern about all of the consequences of mountaintop removal, he also told us that "the most important resource we have is water."

Larry and the other activists we met recognized that water is the essence of life on this planet, where just 0.016 percent of the precious fluid is "active" available freshwater, moving through lakes, rivers, the atmosphere, and living organisms.

We are bodies of water. Humans can live without food for weeks, but die within days when deprived of water. Our bodies are two-thirds water; our brains, more than 95 percent. A 150-pound human body contains more than 12 gallons of water. We must replenish about two and a half quarts each day in drink and food, as we lose water in sweat, urine, and breath. Like every other organism, we have evolved to conserve internal water. Barriers of skin and mucous membranes regulate water passage in and out. Bark and scales serve the same purpose in other organisms. Each living cell has a membrane to enclose and regulate its internal balance of water and essential chemicals, maintaining homeostasis, the internal conditions necessary for life. Water is so essential that it is amazing we ever take it for granted.

Yet, again and again, we saw how a short-term focus on economic gain and resource extraction ignores long-term effects on water environments. Even so, wherever things had been damaged, it was inspiring to see renewal being sought and achieved. Sado Island, in Japan, best illustrates the possibility of restoration and recovery. There, an entire community is working toward the common goal of creating a harmonious water environment for humans and nature, one that will nurture the Japanese Crested Ibis.

Our journeys began at Mono Lake, another place that has become a model of successful recovery against seemingly hopeless odds, and a demonstration of the legal obligation governments have to protect such resources as a public trust. Although not all the water stories on the 38th parallel have happy endings—at least, not yet—we found many reasons for hope. Dedicated people were unwilling to simply accept the unacceptable, even though their challenges seemed daunting.

We found a world of bountiful water environments that justify all that dedication and persistence.

Acknowledgments

We thank Nick and Ryan Carle, and Albert and Doris Broughton, for help with logistics, for house- and car-sitting, and for not worrying too much when we disappeared over the horizon. Thanks also to Krista and Philip Devaul, our hosts in Washington D.C., and to Sally Gaines and Rick Kattelmann for reading the draft manuscript.

Rasa Gustaitis, editor of *California Coast & Ocean,* included a condensed version of the California trek chapter in Autumn 2009 issue of that journal.

And we thank Jenny Wapner and Kim Robinson, our editors at the University of California Press during the years that this book was being conceived, researched, and finally published; also Stacy Eisenstark and Francisco Reinking for their work during production, and Steven Baker for thoughtful and thorough copyediting.

We also acknowledge the contributions of the following individuals and organizations, in order of appearance in the book or during our travels:

KOREA

KFEM, Choony Kim, Yong-un Ma, Hang Jin Lee, Jong-Hak Park, and Naree
Jeong, www.kfem.or.kr

Allen Lampson, Jane Hallinger, Pasadena Sister Cities Committee, www.passcc
.org; Peace Hospital Foundation, Dr. Jung Nam

JungRok An, DMZ Ecological Institute, www.ecodmz.or.kr

Jang Moon Young "Moon", Paju City coordinator and Seung-Pyo Hong, Vice-
Mayor of Paju City, www.paju.go.kr

DMZ Forum, www.dmzforum.org

Se-Yung Jun and Ms. Min-Hwa Yoo, U.C. Press representatives, Information
and Culture Korea, www.ick.co.kr; and Ms. Hae-Young Shon, Kyobo Book
Center import manager

CHINA

Kinder (Jinde) Shu, EDG Earth Expedition (Beijing), www.china-edg.com

Abdul Wayit, Silk Road tour guide, www.kashgartours.com

Deputy Director Li Juanzhang and Shen Kai, engineer, at the Yellow River
Delta (Huanghe Sanjiaozhou) National Nature Reserve

Ran Liping, Green Camel Bell, www.gcbcn.org/en; introduced to us by Xiu
Min Li

Pacific Environment; Zhang Songlin, Geography and Environmental Sciences
Department, Northwest Normal University

Fafu Li, Qinghai Institute of Salt Lakes, http://english.isl.cas.cn/au/bi/200905/
t20090522_3249.html; introduced to us by Robert Jellison, Sierra Nevada
Aquatic Research Laboratory, http://vesr.ucnrs.org/pages/snarlmoreinfo.html

Liu Qiang, Vice-General Director, China Association of National Parks and
Scenic Sites, www.npachina.com

Jim Harris and George Archibald, International Crane Foundation, www.saving
cranes.org

Dr. Rick Kattelmann, for advice and companionship in western China

John and Nancy Walter of Mammoth Lakes, California, and Karen and Paul
Amstutz of Yosemite—intrepid travelers in Asia

TURKMENISTAN

Berkeli Atayev, Chief of International Relations, Marketing and Advertising Department, State Committee of Turkmenistan for Tourism and Sport;
www.tourism-sport.gov.tm; introduced to us by Patrick and Lynn de Frei-
tas, Friends of Great Salt Lake, www.fogsl.org

TURKEY

Ipek Tasli, Initiative to Keep Hasankeyf Alive, www.hasankeyfgirisimi.com;
   Thomas Wenidoppler, Stop Ilisu Campaign, http://m-h-s.org/ilisu
Mustafa Kamal Caddesi, Nemrut Tours from Malatya, www.nemruttours.com
Omar Digirmenci, "Mr. Happy," proprietor of the Liman Hotel, Kusadasi,
   www.limanhotel.com

GREECE

Margaret and John Chaconas, who welcomed us to their house near Athens
Paul Tsaros, Archelon Sea Turtle Rescue Center

SICILY

Anna Giordano, Mediterranean Association for Nature, www.migrazione.it

SPAIN

Antonio Saez, Ranger at the Torrevieja and La Mata lagoons
Javi and Virginia Grijalbo, our naturalist guides across southern Spain; http://
   javiergrijalbo.blogspot.com
German Garrote Alonso, Antonio Rivas, Astrid Vargas, Miguel Angel Simon;
   www.lynxexsitu.es; www.lifelince.org

PORTUGAL

José Martins, Quercus—Associação Nacional de Conservação da Natureza,
   www.quercus.pt

UNITED STATES

Chesapeake Bay

Joelle Buffa, Biologist at the Chincoteague National Wildlife Refuge, and her
   husband, Clyde Morris, for educating us and sharing their home; introduced to us by Emilie Strauss; Myrna Cherrix, Assateague lighthouse historian, www.fws.gov/northeast/chinco
Ann Gilmore, Coastal Virginia Wildlife Observatory, www.cvwo.org

Jack Russell and Krispen Parke, Environmental Educators with the Chesapeake Bay Foundation, www.cbf.org

John Tippett, Friends of Rappahannock, www.riverfriends.org

Jim Long and Bonnie Bick, Mattawoman Watershed Society, http://mattawoman watershedsociety.org/

West Virginia

Justin Reynolds, Guide, and Skip Heater, owner, New Gauley River Adventure, www.gauley.com

Larry Gibson, Keeper of the Mountain Foundation, http://mountainkeeper.blog spot.com/

Roland Micklem and Andrew Munn, Seniors March to End Mountaintop Removal

Kentucky

Malcolm and Judy Dalgleish, who hosted us in Indiana and introduced us to Wendell and Tanya Berry

Akiko McVarish Gothard, Heatherway, equine management

Jim and Kay Boudinot, for introducing us to Akiko

Bev and John Passarello, Passarello Thoroughbreds

Larry Meadows, Red River Museum

Indiana

Warren and Jill Schimpff, Schimpff's Candy Store

Missouri

Terry Cunningham, Jason Green, and Brandon Kuhn of the Pioneer Forest, www.pioneerforest.org

Kansas

Eric Patterson, Tallgrass Prairie National Preserve, www.nps.gov/tapr

Rachel Laubhan and Ranger Barry Jones, Quivira National Wildlife Refuge, www.fws.gov/quivira

Mark Rude and Jason Norquist, Southwest Kansas Groundwater Management District No. 3, www.gmd3.org

Nancy Harness, Mayor of Garden City, www.garden-city.org; introduced to us by Molly DesBaillets

Mark Sexson, Kansas Department of Wildlife and Parks, www.kdwp.state.ks
.us
Tom Norman, Sandsage Bison Range, www.kdwp.state.ks.us/newsKDWP
-Info/Locations/Wildlife-Areas/Region-3/Sandsage-Bison-Range

Colorado

Shannon Nelson and Brett Pyle, our hosts on a snowy night in Salida
Vicki Klein, Betsy Steinberg, and John and Ann Graham, Chafee County
    Citizens for Sustainability, http://ccfsustainability.org
Paul Robertson, The Nature Conservancy, www.nature.org
Walter Wright, Telluride 350, www.climategroundzero.org

Utah

Elizabeth "Liz" Glowiak and Ken Pill, Moab TAC Team, and Donald Metzler,
    U.S. Department of Energy, www.gjem.energy.gov/moab
Kimberly Schappert, Up the Creek Campground, Moab Trails Alliance
"Bodie Jack" Shipley, our Green River guide

Nevada

Glennon Zelch, Pioche

California

Geoff McQuilkin, Mono Lake Committee, www.monolake.org
Connie Millar, U.S. Forest Service, Pacific Southwest Research Station, www.fs
    .fed.us/psw
Spreck Rosekrans, Restore Hetch Hetchy, www.hetchhetchy.org
Kurt Stegen, Friends of the River, www.friendsoftheriver.org
John Knott, David Martin, Marty Knott, and Skippy the dog—sailboat captains
    and crew
John Cain, Natural Heritage Institute, www.n-h-i.org
Marc Holmes, The Bay Institute, www.bay.org
Stan and Sharon Bluhm, Mono Lake volunteers who opened their house to us
    in San Anselmo; Richard and Brenda Nichols, Coastwalk California, www
    .coastwalk.org
Don Neubacher, Point Reyes National Seashore, www.nps.gov/pore
Allen Fish, Golden Gate Raptor Observatory, www.ggro.org
David Wimpfheimer (www.calnaturalist.com), our guide for the Marin
    Audubon Society trip to Farallones, http://marinaudubon.aviandesign.net

JAPAN

Tsuneo Watanabe, Director General, Nature Conservation Bureau, Ministry of
the Environment; Tetsuro Uesugi, Director, National Park Division; Shin-
jiro Sasaki, Keisuke Takahashi, and Kohei Sasabuchi, Assistant Directors,
National Park Division; Hiroshi Horiuchi, Wildlife Division
Tsuneo Sekijima, Department of Environmental Science, Niigata University; Dr.
Chihiro Endo, Toki and Nature Restoration Research Center, Sado Island;
Taeko Ishima, graduate student at Niigata University
Yoichi Kawaguchi, Tokushima University
Kei Osada, Ranger, Crested Ibis Conservancy Center, Sado Island, www.visit
sado.com/en/00sp/0809/toki1.shtml
Akira Wada, Asian River Restoration Network, www.a-rr.net

Bibliography

Ashworth, William. 2006. *Ogallala blue: Water and life on the high plains.* New York: W. W. Norton.

Baumer, Christoph. 2003. *Southern Silk Road: In the footsteps of Sir Aurel Stein and Sven Hedin.* Bangkok: Orchid Press.

Bazairi, H., S. Ben Haj, F. Boero, D. Cebrian, S. De Juan, A. Limam, J. Lleonart, G. Torchia, and C. Rais, eds. 2010. "The Mediterranean Sea biodiversity: State of the ecosystems, pressures, impacts and future priorities." Tunis: United Nations Environment Programme, Mediterranean Action Plan, RAC/SPA.

Berry, Wendell. 1991. *The unforeseen wilderness: Kentucky's Red River Gorge.* Emeryville, CA: Shoemaker Hoard.

———. 1998. *The selected poems of Wendell Berry.* Berkeley, CA: Counterpoint Press.

———. 2002. *The art of the commonplace.* Berkeley, CA: Counterpoint Press.

Bonavia, Judy. 2008 [1998]. *The silk road: Xian to Kashgar.* Hong Kong: Odyssey.

Brown, Lester R. 2009. *Plan B 4.0: Mobilizing to save civilization.* New York: W. W. Norton.

Buck, Pearl S. 1963 [1931]. *The good earth*. New York: Cardinal/Pocket Books.

Burdick, Charles F. 2003. *Stories of river pilots on the great western rivers*. Grand Tower, IL: BurPublishing.

Carle, David. 2003. *Water and the California dream*. San Francisco: Sierra Club Books.

———. 2009 [2004]. *Introduction to water in California*. Berkeley: University of California Press.

California Department of Water Resources. 2006. "Hetch Hetchy Restoration Study." http://hetchhetchy.water.ca.gov/docs/Hetch_Hetchy_Restoration_ Study_Report.pdf (accessed December 6, 2010).

Carle, David, with Janet Carle. 2009. "Exploring California's 'Water Line.'" *California Coast & Ocean* 25(2): 11–20.

Chamberlin, J. Edward. 2006. *Horse: How the horse has shaped civilizations*. New York: BlueBridge.

Cherniack, Martin. 1986. *The Hawk's Nest incident: America's worst industrial disaster*. New Haven: Yale University Press.

CIA. 2012. "The World Factbook." Central Intelligence Agency. www.cia .gov/library/publications/the-world-factbook/rankorder/2127rank.html (accessed March 26, 2012).

Clarke, Thurston. 1988. *Equator: A journey*. New York: Avon Books.

Colombo, Francesca. 2004. "Mafia kidnapping Sicily's water." Inter Press Service. *Albion Monitor*. June 28. www.monitor.net/monitor/0406a/copyright/ mafiawater.html (accessed March 20, 2012).

Dai Qing. 1998. *The river dragon has come! The Three Gorges Dam and the fate of China's Yangtze River and its people*. International Rivers Network. Armonk, NY: M.E. Sharpe.

Economy, Elizabeth C. 2010 [2004]. *The river runs black: The environmental challenge to China's future*. Ithaca, NY: Cornell University Press.

Egleston, N.H. 1909. "Granting use of Hetch Hetchy to City of San Francisco." Report to 60th Congress. February 8. Washington, D.C.

Fagan, Brian. 2011. *Elixir: A history of water and humankind*. New York: Bloomsbury Press.

Fatakuchi, Kazuko. 2010. "Transforming agriculture and economy to save the Japanese Crested Ibis: Sado Island." JFS Newsletter, Japan Ministry of the Environment. No. 97 (September 2010). www.japanfs.org/en/ mailmagazine/newsletter/pages/030310.html (accessed March 4, 2011).

Geredieu, Ross. 2009. "Update: Extent of mountaintop mining in Appalachia as of 2008." Boone, NC: Appalachian Voices. www.ilovemountains.org/ reclamation-fail/mining-extent-2009/MTR_Peaks_Update.pdf (accessed February 20, 2010).

Giglioli, Ilaria, and Erik Swyngedouw. 2008. "Let's drink to the great thirst! Water and the politics of fractured techno-natures in Sicily." *International Journal of Urban and Regional Research* 32(2): 392–414 (online edition accessed March 20, 2012).

Gow, A. S. F., and Page, D. L. (editors). 1968. *The Greek anthology: The garland of Philip and some contemporary epigrams.* Cambridge: Cambridge University Press.

Hessler, Peter. 2002. *River town: Two years on the Yangtze.* New York: Perennial/ Harper Collins.

Horton, Tom. 1991. *Turning the tide: Saving the Chesapeake Bay.* Washington, DC: Chesapeake Bay Foundation; Island Press.

———. 1996. *An island out of time: A memoir of Smith Island in the Chesapeake.* New York: W. W. Norton.

Irving, Washington. 1994. *Tales of the Alhambra.* Granada, Spain: Ediciones Miguel Sanchez.

"It's a Hoax, All Right." 1974. Editorial. *Los Angeles Times.* October 25. Pt. 2, p. 6.

Ivanova, Nadya. 2011. "Water rights transfers and high-tech power plants hold off energy-water clash in northern China." Circle of Blue. March 8. www .circleofblue.org/waternews/2011/world/from-agriculture-to-industry -efficiency-upgrades-transfer-water-use-rights-on-china (accessed March 9, 2011).

Jong-hwi, Jeon. 2010. "Government to increase the number of protected areas within the DMZ." *The Hankyoreh.* September 3. www.hani.co.kr/arti/ english_edition/e_national/438155.html (accessed September 25, 2010).

Kreutzmann, Hermann. 2003. "Ethnic minorities and marginality in the Pamirian Knot: Survival of Wakhi and Kirghiz in a harsh environment and global contexts." *Geographical Journal* 169(3) (September): 215–35.

Larmer, Brook. 2008. "Bitter waters: Can China save the Yellow, its mother river?" *National Geographic* (May): 146–69.

Least Heat-Moon, William. 1999. *PrairyErth (a deep map): An epic history of the tallgrass prairie country.* Boston: Mariner Books.

Malloy, Daniel. 2011. "EPA vetoes mining permit in West Virginia." *Pittsburgh Post-Gazette.* January 14. www.post-gazette.com/pg/11014/1117862–113.stm (accessed February 22, 2011).

Matthiessen, Peter. 2001. *The birds of Heaven: Travels with cranes.* New York: North Point Press.

Mertha, Andrew C. 2008. *China's water warriors: Citizen action and policy change.* Ithaca, NY: Cornell University Press.

Mount, Jeffrey F. 1995. *California rivers and streams.* Berkeley: University of California Press.

Muir, John. 1912. *The Yosemite.* New York: Century Company.

Nakamura, Keigo, Klement Tockner, and Kunihiko Amano. 2006. "River and wetland restoration: Lessons from Japan." *BioScience* 56(5) (May): 419–29.

Nash, Ogden. 1931. *Free wheeling.* New York: Simon and Schuster.

Nation, Allan H. 2010. "Allan's observations." *Stockman Grass Farmer* 10(1).

National Research Council. 1981. *Effect of environment on nutrient requirements of domestic animals.* Washington, DC: National Academies Press.

Ohio River Foundation. 2012. "Water quality protection: Regulatory oversight showing weaknesses." www.ohioriverfdn.org/stewardship/water_quality_protection/ (accessed March 23, 2012).

Opie, John. 2000. *Ogallala: Water for a dry land.* Lincoln: University of Nebraska Press.

Pacific Institute. N.d. "Bottled water and energy: a fact sheet." www.pacinst.org/topics/water_and_sustainability/bottled_water/bottled_water_and_energy.html (accessed February 25, 2010).

Palmer, Tim. 1982. *Stanislaus: The struggle for a river.* Berkeley: University of California Press.

Paul, Kathleen. 2011. "The story behind the fund." Matsushima Relief Fund. https://sites.google.com/site/matsushimafund/background (accessed September 19, 2011).

Polo, Marco. 1982 [1958]. *The travels of Marco Polo.* Translated and introduced by Ronald Latham. New York: Abaris Books.

Powell, John Wesley. 1879. *Report on the lands of the arid region of the United States.* Washington, DC: U.S. Government Printing Office.

Prosek, James. 2003. *Fly-fishing the 41st: Around the world on the 41st parallel.* New York: HarperCollins.

Ramzay, Austin. 2009. "Heroes of the environment, 2009: Zhao Zhong." *Time.* September 22.

Righter, Robert W. 2005. *The battle over Hetch Hetchy: America's most controversial dam and the birth of modern environmentalism.* New York: Oxford University Press.

Rosenburg, Matt. T. 2011. "Largest Cities through History." About.com Geography. http://geography.about.com/library/weekly/aa011201a.htm (accessed October 8, 2011).

Schneider, Keith. 2011. "Choke point: China—Confronting water scarcity and energy demand in the world's largest country." Circle of Blue. February 15. www.circleofblue.org/waternews/2011/world/choke-point-chinaconfront

ing-water-scarcity-and-energy-demand-in-the-world's-largest-country/ (accessed February 22, 2011).

Sekijima, Tsuneo, Yoichi Kawaguchi, Tadashi Miyashita, Yasuhiro Mitani, and Ryugo Watanabe. 2010. *Messages from Japan's green pioneers: Living in harmony with nature.* Translated into English from *Environmental Research Quarterly* by the Hitachi Environment Foundation. Tokyo: Ministry of the Environment, Government of Japan.

Shea, Kelly. 2011. "Circle of Blue Infographic: Going the distance, from Ashgabat to Whyalla—10 cities pumping water from afar." Circle of Blue. February 1. www.circleofblue.org/waternews/2011/world/infographic-going-the-distance/ (accessed February 3, 2011).

Simonds, Wm. Joe. 1994. "New Melones unit project history." U.S. Bureau of Reclamation. http://www.usbr.gov/projects (accessed February 29, 2010).

Smith, Andrew T., and J. Marc Foggin. 1999. "The plateau pika (*Ochotona curzoniae*) is a keystone species for biodiversity on the Tibetan Plateau." *Animal Conservation* 2: 235–40.

Spangler, Patricia. 2008. *The Hawks Nest tunnel: An unabridged history.* Proctorville, OH: Wythe-North.

Takano, Koichiro. 2010. "Declaring 'biological survey days.'" In *Sado City's approach to economic and environmental synergies.* Sado City, Niigata Prefecture, Japan.

Tiller & Toiler. 2009. "Fort Larned 150th anniversary, 1859–2009." Special Commemorative Section. October 11. Larned, Kansas.

Twain, Mark. 1873. *Roughing it.* Hartford, CT: American Publishing Co.

U.S. Environmental Protection Agency. 2009. "The effects of mountaintop mines and valley fills on aquatic ecosystems of the Central Appalachian coalfields." DRAFT. EPA/600/R-09/138A. Washington, DC: Office of Research and Development, National Center for Environmental Assessment.

U.S. Geological Survey. 2011. "Poster of the Great Tohoku Earthquake (northeast Honshu, Japan) of March 11, 2011—Magnitude 9.0." http://earthquake.usgs.gov/earthquakes/eqarchives/poster/2011/20110311.php (accessed September 27, 2011).

U.S. National Park Service. 1988. "Alternatives for restoration of Hetch Hetchy Valley following removal of the dam and the reservoir." www.sierraclub.org/ca/hetchhetchy/nps_hh_restoration.pdf (accessed December 10, 2010).

Vervaeck, Armand, and James Daniell. 2011. "CATDAT Sendai Japan total situation report v37." Earthquake Report.com. http://earthquake-report.com/2011/08/04/japan-tsunami-following-up-the-aftermath-part-16-june/ (accessed October 1, 2011).

Ward, Ken, Jr. 2009. "Byrd blasts Massey 'arrogance' over school." *Charleston Gazette*. October 8: 1.

Water Environment Department. 2012. "History of water environment policy." In *Water Environmental Management in Japan*. Tokyo: Environmental Management Bureau, Ministry of the Environment. www.env.go.jp/en/water/wq/pamph/ (accessed March 22, 2012).

Xinhua News Agency. 2008. "Massive program launched to save Qinghai Lake." *China Daily*. May 26. www.chinadaily.com.cn/china/2008–05/26/ (accessed February 6, 2010).

Yu, Junqing, and Lisa Zhang. 2008. *Lake Qinghai: Paleoenvironment and paleoclimate*. Beijing: Science Press.

INDEX